AF686620

SOIL SCIENCE: PRINCIPLES AND TECHNOLOGY

Edited by
Taras Kazantsev

ArclerPress

Soil Science: Principles and Technology

Edited by Taras Kazantsev

© 2017

Edition 2018

ArclerPress

Published by
Arcler Press LLC
708 3rd Avenue, 6th Floor
New York
NY 10017
United States of America
www.arclerpress.com

ISBN: 978-1-68094-562-1 (Hardback)

Printed in United States of America

Library of Congress Control Number: 2017933132

This book contains information obtained from highly regarded resources. Some content has been published with permission under Creative Commons License or similar. A Wide variety of references are listed. Reasonable efforts have been made to publish reliable data and views articulated in the chapters are those of the individual contributors, and not necessarily those of the editors or publishers. Editors or publishers are not responsible for the accuracy of the information in the published chapters or consequences of their use. The publisher believes no responsibility for any damage or grievance to the persons or property arising out of the use of any materials, instructions, methods or thoughts in the book. The editors and the publisher have attempted to trace the copyright holders of all material reproduced in this publication and apologize to copyright holders if permission has not been obtained. If any copyright holder has not been acknowledged, please write to us so we may rectify.

Notice: Registered trademark of products or corporate names are used only for explanation and identification without intent of infringement. Although care has been taken to check accuracy of formulas and procedures, the detailed methods should be tested further on a small scale before being adopted commercially.

Arcler Press publishes wide variety of books and eBooks. For more information about Arcler Press and its products, visit our website at www.arclerpress.com.

About the Editor

Taras Kazantsev

Taras Kazantsev has PhD in plant physiology since 2008. He is expert in remote sensing of vegetation and aerial mapping. He is authors of 15 papers and 2 patents. Besides research activity and articles writing, Taras Kazantsev is involved in a number of commercial projects on agromonitoring. Currently, Taras Kazantsev is research associate in Centre for Aero-Space research of Earth in Kyiv, Ukraine and head of Remote Sensing department in private company Drone.UA

Table of Contents

3 SOIL FORMING AND DEGRADATION PROCESSES

4 SOIL CHEMISTRY

5 SOIL SALINITY

6 SOIL CLASSIFICATION

7 ECOLOGICAL STUDIES OF SOIL

8 SOIL BIOLOGY

9 SOIL CONSERVATION AND MANAGEMENT

10 LAND MANAGEMENT

11 SOIL GOVERNANCE

List of Abbreviations

Al	Aluminium
AEC	Anion exchange capacity
C/N	Carbon to nitrogen ratio
CEC	Cation exchange capacity
CEC	Cation-exchange capacity
C	Cultivation
ESP	Exchangeable sodium percentage
FAO	Food and Agriculture Organization
GIS	Geographic Information Systems
GSP	Global Soil Partnership
ITPS	Intergovernmental Technical Panel of Soils
Mn	Manganese
NO	Nitric oxide
RSPs	Regional Soil Partnerships

Preface

The food that nourishes us, the water that we drink and the air that we breathe all depend on soils and their life-sustaining properties. Soil Science provides an understanding of the practical application of biology, chemistry, physics and earth science principles to integrated land use and environmental protection. It deals with soil as a natural resource on the surface of the earth including soil formation, classification and mapping; physical, chemical, biological, and fertility properties of soils per se; and these properties in relation to the use and management of soils. Oil is a renewable energy source and is one of the 'life-bloods' that nurtures, nourishes and sustains life. Of the total food consumed by humans, 99.7% comes from soil. It is estimated that the world wastes between 80-90% of its applied nutrients and yet soil health, its ability to operate as a carbon sink, and its role in revitalising waterways are vital priorities for industry and shaping agrifood practices. A soil scientist studies the chemical, physical, and biological properties of the top few meters of the Earth's surface. Soil science is divided into two main areas of study - soil pedology and soil edaphology. Soil pedology studies the formation and development, composition, morphology and identification and the classification of soils. Edaphology focuses on human's use of soil and its influence on living things, particularly plants. This includes soil fertility, managing sewers and landfills, anticipating flood runoff, and examining the process by which plants obtain nutrients and water from soil. Scientists also look at the inter-relationships between soil and factors that affect plant growth such as soil acidity, salinity and toxic contaminations.

Content Coverage

Chapter One is intended to provide an overview of history of study of soil, including history and notions relating to crop ecology.

Chapter Two aims to provide a comprehensive overview of physical, chemical, biological, and fertility properties of soils; and these properties in relation to the use and management of soils.

Chapter Three discusses about the soil formation and degradation processes. The formation of a new soil depends on the presence of new soil material - obtained either by denudation - the scraping clean of a surface by action of wind, water or ice - or deposition of new materials from erosion elsewhere.

Chapter Four reviews the theory and principles of soil chemistry that deals with study of the chemical characteristics of soil. Soil chemistry is affected by mineral composition, organic matter and environmental factors.

Chapters Five and *Six* present about soil salinity control and soil classification, separately. The purpose of soil salinity control is to prevent soil degradation by salinization and reclaim already salty soils. Soil classification is the separation of soil into classes or groups each having similar characteristics and potentially similar behavior.

Chapters Seven and *Eight* provide the studies of soil ecology and soil biology, respectively. Soil ecology is the study of the interactions among soil organisms, and between biotic and abiotic aspects of the soil environment, while soil biology deals with the study of microbial and faunal activity and ecology in soil.

Chapters Nine and *Ten* provide the knowledge of soil conservations and land management, respectively. Soil conservation is the prevention of soil loss from erosion or reduced fertility caused by over usage, acidification, salinization or other chemical soil contamination. Land management practices have become an increasing focus of the soil science that involves the management and modification of natural environment or wilderness.

Chapter Eleven is soil governance focused to the policies, strategies, and the processes of decision-making employed by nation states and local governments regarding the use of soil.

CHAPTER 1

History of Study of Soil

Chapter Contents

- **Studies Concerning Soil Fertility**
- **Prominent Soil Physicists**
- **History of Agroecology**

Soil is not only a support for vegetation, but it is also the zone beneath our feet (the pedosphere) of numerous interactions between climate (water, air, temperature), soil life (micro-organisms, plants, animals) and its residues, the mineral material of the original and added rock, and its position in the landscape. During its formation and genesis, the soil profile slowly deepens and develops characteristic layers, called 'horizons', while a steady state balance is approached.

Soil users (such as agronomists) showed initially little concern in the dynamics of soil. They saw it as medium whose chemical, physical and biological properties were useful for the services of agronomic productivity . On the other hand, pedologists and geologists did not initially focus on the agronomic applications of the soil characteristics (edaphic properties) but upon its relation to the nature and history of landscapes. Today, there's an integration of the two disciplinary approaches as part of landscape and environmental sciences.

Pedologists are now also interested in the practical applications of a good understanding of pedogenesis processes (the evolution and functioning of soils), like interpreting its environmental history and predicting consequences of changes in land use, while agronomists understand that the cultivated soil is a complex medium, often resulting from several thousands of years of evolution. They understand that the current balance is fragile and that only a thorough knowledge of its history makes it possible to ensure its sustainable use.

STUDIES CONCERNING SOIL FERTILITY

The history of the study of soil is intimately tied to our urgent need to provide food for ourselves and forage for our animals. Throughout history, civilizations have prospered or declined as a function of the availability and productivity of their soils.

The Greek historian Xenophon (450–355 B.C.) is credited with being the first to expound upon the merits of green-manuring crops: "But then whatever weeds are upon the ground, being turned into earth, enrich the soil as much as dung."

Columella's "Husbandry," circa 60 A.D., advocated the use of lime and that clover and alfalfa (green manure) should be turned under, and was used by 15 generations (450 years) under the Roman Empire until its collapse. From the fall of Rome to the French Revolution, knowledge of soil and agriculture was passed on from parent to child and as a result, crop yields were low. During the European Dark Ages, Yahya Ibn al-'Awwam's handbook, with its emphasis on irrigation, guided the people of North Africa, Spain and the Middle East; a translation of this work was finally carried to the southwest of the United States.

Experiments into what made plants grow first led to the idea that the ash left behind when plant matter was burned was the essential element but overlooked the role of nitrogen, which is not left on the ground after combustion. In about 1635, the Flemish chemist Jan Baptist van Helmont thought he had proved water to be the essential element from his famous five years' experiment with a willow tree grown with only the addition of rainwater. His conclusion came from the fact that the increase in the plant's weight had apparently been produced only by the addition of water, with no reduction in the soil's weight. John Woodward (d. 1728) experimented with various types of water ranging from clean to muddy and found muddy water the best, and so he concluded that earthy matter was the essential element. Others concluded it was humus in the soil that passed some essence to the growing plant. Still others held that the vital growth principal was something passed from dead plants or animals to the new plants. At the start of the 18th century, Jethro Tull demonstrated that it was beneficial to cultivate (stir) the soil, but his opinion that the stirring made the fine parts of soil available for plant absorption was erroneous.

As chemistry developed, it was applied to the investigation of soil fertility. The French chemist Antoine Lavoisier showed in about 1778 that plants and animals must "combust" oxygen internally to live and was able to deduce that most of the 165-pound weight of van Helmont's willow tree derived from air. It was the French agriculturalist Jean-Baptiste Boussingault who by means of experimentation obtained evidence showing that the main sources of carbon, hydrogen and oxygen for plants were the air and water. Justus von Liebig in his book *Organic Chemistry in its Applications to Agriculture and Physiology* (published 1840), asserted that the chemicals in plants must have come from the soil and air and that to maintain soil fertility, the used minerals must be replaced. Liebig nevertheless believed the nitrogen was supplied from the air. The enrichment of soil with guano by the Incas was rediscovered in 1802, by Alexander von Humboldt. This led to its mining and that of Chilean nitrate and to its application to soil in the United States and Europe after 1840. The work of Liebig was a revolution for agriculture, and so other investigators started experimentation based on it. In England John Bennet Lawes and Joseph Henry Gilbert worked in the Rothamsted Experimental Station, founded by the former, and discovered that plants took nitrogen from the soil, and that salts needed to be in an available state to be absorbed by plants. Their investigations also produced the «superphosphate», consisting in the acid treatment of phosphate rock.This led to the invention and use of salts of potassium (K) and nitrogen (N) as fertilizers. Ammonia generated by the production of coke was recovered and used as fertiliser. Finally, the chemical basis of nutrients delivered to the soil in manure was understood and in the mid-19th century chemical fertilisers were applied. However, the dynamic interaction of soil and its life forms awaited discovery.

In 1856 J. T. Way discovered that ammonia contained in fertilisers was transformed into nitrates, and twenty years later R. W. Warington proved that this transformation was done by living organisms. In 1890 Sergei Winogradsky announced he had found the bacteria responsible for this transformation.

It was known that certain legumes could take up nitrogen from the air and fix it to the soil but it took the development of bacteriology towards the end of the 19th century to lead to an understanding of the role played in nitrogen fixation by bacteria. The symbiosis of bacteria and leguminous roots, and the fixation of nitrogen by the bacteria, were simultaneously discovered by German agronomist Hermann Hellriegel and Dutch microbiologist Martinus Beijerinck.

Crop rotation, mechanisation, chemical and natural fertilisers led to a doubling of wheat yields in Western Europe between 1800 and 1900.

Studies Concerning Soil Formation

The scientists who studied the soil in connection with agricultural practices had considered it mainly as a static substrate. However, soil is the result of evolution from more ancient geological materials. After studies of the improvement of the soil commenced, others began to study soil genesis and as a result also soil types and classifications.

In 1860, in Mississippi, Eugene W. Hilgard studied the relationship among rock material, climate, and vegetation, and the type of soils that were developed. He realised that the soils were dynamic, and considered soil types classification. Unfortunately his work was not continued. At the same time Vasily Dokuchaev (about 1870) was leading a team of soil scientists in Russia who conducted an extensive survey of soils, finding that similar basic rocks, climate and vegetation types lead to similar soil layering and types, and established the concepts for soil classifications. Due to the language barriers, the work of this team was not communicated to Western Europe until 1914 by a publication in German by K. D. Glinka, a member of the Russian team.

Curtis F. Marbut was influenced by the work of the Russian team, translated Glinka's publication into English, and as he was placed in charge of the U. S. National Cooperative Soil Survey, applied it to a national soil classification system.

PROMINENT SOIL PHYSICISTS

Edgar Buckingham (1867–1940)

The theory of gas diffusion in soil and vadose zone water flow in soil.

Lorenzo A. Richards (1904–1993)

General transport of water in unsaturated soil, measurement of soil water potential using tensiometer.

John R. Philip (1927–1999)

Analytical solution to general soil water transport, Environmental Mechanics.

HISTORY OF AGROECOLOGY

Pre-WWII

The notions and ideas relating to crop ecology have been around since at least 1911 when F.H. King released *Farmers of Forty Centuries*. King was one of the pioneers as a proponent of more quantitative methods for characterization of water relations and physical properties of soils. In the late 1920s the attempt to merge agronomy and ecology was born with the development of the field of crop ecology. Crop ecology's main concern was where crops would be best grown.Actually, it was only in 1928 that agronomy and ecology were formally linked by Klages.

The first mention of the term agroecology was in 1928, with the publication of the term by Bensin in 1928. The book of Tischler (1965), was probably the first to be actually titled 'agroecology'. He analysed the different components (plants, animals, soils and climate) and their interactions within an agroecosystem as well as the impact of human agricultural management on these components. Other books dealing with agroecology, but without using the term explicitly were published by the German zoologist Friederichs (1930) with his book on agricultural zoology and related ecological/environmental factors for plant protection, and by American crop physiologist Hansen in 1939 when both used the word as a synonym for the application of ecology within agriculture.

Post-WWII

Gliessman mentions that post-WWII, groups of scientists with ecologists gave more focus to experiments in the natural environment, while agronomists dedicated their attention to the cultivated systems in agriculture. According to Gliessman, the two groups kept their research and interest apart until books and articles using the concept of agroecosystems and the word agroecology started to appear in 1970. Dalgaard explains the different points of view in ecology schools, and the fundamental differences, which set the basis for the development of agroecology. The early ecology school of Henry Gleason investigated plant populations focusing in the hierarchical levels of the organism under study.

Friederich Clement's ecology school, however included the organism in question as well as the higher hierarchical levels in its investigations, a "landscape perspective". However, the ecological schools where the roots of agroecology lie are even broader in nature. The ecology school of Tansley, whose view included both the biotic organism and their environment, is the one from which the concept of agroecosystems emerged in 1974 with Harper. In the 1960s and 1970s the increasing awareness of how humans manage the landscape and its consequences set the stage for the necessary cross between agronomy and ecology. Even though, in many ways the environmental movement in the US was a product of the times, the Green Decade, spread an environmental awareness of the unintended consequences of changing ecological processes. Works such as *Silent Spring*, and *The Limits to Growth*, and changes in legislation such as the Clean Air Act, Clean Water Act, and the National Environmental Policy Act caused the public to be aware of societal growth patterns, agricultural production, and the overall capacity of the system.

Fusion with Ecology

After the 1970s, when agronomists saw the value of ecology and ecologists began to use the agricultural systems as study plots, studies in agroecology grew more rapidly. Gliessman describes that the innovative work of Prof. Efraim Hernandez X., who developed research based on indigenous systems of knowledge in Mexico, led to education programs in agroecology. In 1977 Prof. Efraim Hernandez X. explained that modern agricultural systems had lost their ecological foundation when socio-economic factors became the only driving force in the food system. The acknowledgement that the socio-economic interactions are indeed one of the fundamental components of any agroecosystems came to light in 1982, with the article Agroecologia del Tropico Americano by Montaldo. The author argues that the socio-economic context cannot be separated from the agricultural systems when designing agricultural practices.

In 1995 Edens et al. in Sustainable Agriculture and Integrated Farming Systems solidified this idea proving his point by devoting special sections to economics of the systems, ecological impacts, and ethics and values in agriculture. Actually, 1985 ended up being a fertile and creative year for the new discipline. For instance in the same year, Miguel Altieri integrated how consolidation of the farms, and cropping systems impact pest populations. In addition, Gliessman highlighted that socio-economic, technological, and ecological components give rise to producer choices of food production systems. These pioneering agroecologists have helped to frame the foundation of what we today consider the interdisciplinary field of agroecology and have led to advances in a number of farming systems. In Asian rice, for example, crop diversification by growing flowering crops in strips beside rice fields has recently been demonstrated to reduce pests so effectively (by the flower nectar attracting and supporting parasitoids and predators) that insecticide spraying is reduced by 70%, yields increase by 5%, together resulting in an economic advantage of 7.5% (Gurr et al., 2016).

Publications

Year	Author(s)	Title
1928	Klages	Crop ecology and ecological crop geography in the agronomic curriculum
1939	Hanson	Ecology in agriculture

1956	Azzi	Agricultural ecology
1965	Tischler	Agrarökologie
1973	Janzen	Tropical agroecosystems
1974	Harper	The need for a focus on agro-ecosystems
1976	Loucks	Emergence of research on agroecosystems
1977	Hernandez Xolocotzi	Agroecosistemas de Mexico
1978	Gliessman	Agroecosistemas y tecnologia agricola tradicional
1979	Hart	Agroecosistemas: conceptos básicos
1979	Cox & Atkins	Agricultural ecology: an analysis of world food production systems
1980	Hart	Agroecosistemas
1981	Gliessman, Garcia & Amador	The ecological basis for the application of traditional agricultural technology in the management of tropical agroecosystems
1982	Montaldo	Agroecologia del trópico americano
1983	Altieri	Agroecology
1984	Lowrance, Stinner & House	Agricultural ecosystems: unifying concepts
1985	Conway	Agroecosystems analysis
1987	Altieri	Agroecology: the scientific basis of alternative agriculture
1990	Allen, Dusen, Lundy, & Gliessman	Integrating social, environmental, and economic issues in sustainable agriculture
1990	Gliessman	Agroecology: researching the ecological basis for sustainable agriculture
1990	Carroll, Vandermeer & Rosset	Agroecology
1990	Altieri & Hecht	Agroecology and small farm development
1991	Caporali	Ecologia per l'agricultura
1991	Bawden	Systems thinking in agriculture
1993	Coscia	Agricultura sostenible
1998	Gliessman	Agroecology: ecological processes in sustainable agriculture
2001	Flora	Interactions between agroecosystems and rural communities
2001	Gliessman	Agroecosystem sustainability
2002	Dalgaard, Porter & Hutchings	Agroecology, scaling, and interdisciplinarity

2003	Francis et al.	Agroecology: The Ecology of Food Systems
2004	Clements, Shrestra	New Dimension in Agroecology
2007	Bland and Bell	A Holon Approach to Agroecology
2007	Gliessman	Agroecology: The Ecology of Sustainable Food Systems
2007	Warner	Agroecology in Action
2009	Wezel, Soldat	A quantitative and qualitative historical analysis of the scientific discipline agroecology
2009	Wezel et al.	Agroecology as a science, a movement or a practice. A review

By Region

The principles of agroecology are expressed differently depending on local ecological and social contexts.

Latin America

Latin America's experiences with North American Green Revolution agricultural techniques have opened space for agroecologists. Traditional or indigenous knowledge represents a wealth of possibility for agroecologists, including "exchange of wisdoms". See Miguel Alteiri's *Enhancing the Productivity of Latin American Traditional Peasant Farming Systems Through an Agroecological Approach* for information on agroecology in Latin America.

Madagascar

Most of the historical farming in Madagascar has been conducted by indigenous peoples. The French colonial period disturbed a very small percentage of land area, and even included some useful experiments in Sustainable forestry. Slash-and-burn techniques, a component of some shifting cultivation systems have been practised by natives in Madagascar for centuries. As of 2006 some of the major agricultural products from slash-and-burn methods are wood, charcoal and grass for Zebu grazing. These practices have taken perhaps the greatest toll on land fertility since the end of French rule, mainly due to overpopulation pressures.

REFERENCES

1. Voroney, R. Paul; Heck, Richard J. (2015). "The soil habitat". In Paul, Eldor A. Soil microbiology, ecology and biochemistry (4th ed.). Amsterdam, The Netherlands: Elsevier. doi:10.1016/B978-0-12-415955-6.00002-5. ISBN 978-0-12-415955-6.
2. Danoff-Burg, James A. "The terrestrial influence: geology and soils". Earth

Institute Center for Environmental Sustainability. New York, New York: Columbia University Press. Retrieved 25 September 2016.

3. Taylor, Sterling A.; Ashcroft, Gaylen L. (1972). Physical edaphology: the physics of irrigated and nonirrigated soils. San Francisco, California: W.H. Freeman. ISBN 978-0716708186.
4. McCarthy, David F. (2006). Essentials of soil mechanics and foundations: basic geotechnics (7th ed.). Upper Saddle River, New Jersey: Prentice Hall. ISBN 978-0131145603.
5. Gilluly, James; Waters, Aaron Clement; Woodford, Alfred Oswald (1975). Principles of geology (4th ed.). San Francisco, California: W.H. Freeman. ISBN 978-0716702696.
6. Ponge, Jean-François (2015). "The soil as an ecosystem" (PDF). Biology and Fertility of Soils. 51 (6): 645–648. doi:10.1007/s00374-015-1016-1.
7. "Soil density". Argonne National Laboratory. Retrieved 25 September 2016.
8. Buol, Stanley W.; Southard, Randal J.; Graham, Robert C.; McDaniel, Paul A. (2011). Soil genesis and classification (6th ed.). Ames, Iowa: Wiley-Blackwell. ISBN 978-0470960608.
9. Retallack, Gregory J. (2008). Soils of the past: an introduction to paleopedology (2nd ed.). Hoboken, New Jersey: John Wiley & Sons. ISBN 978-0470698167.
10. "Glossary of Terms in Soil Science". Agriculture and Agri-Food Canada. Retrieved 9 October 2016.
11. Amundson, Ronald. "Soil preservation and the future of pedology" (PDF). Faculty of Natural Resources. Prince of Songkla University, Songkhla, Thailand. Retrieved 9 October 2016.
12. Simonson, Roy W. (1957). "What soils are". The yearbook of agriculture 1957(PDF) (1st ed.). Washington, D.C.: United States Government Printing Office.
13. Raloff, Janet (17 July 2008). "Dirt is not soil". ScienceNews. Retrieved 23 October 2016.
14. Pouyat, Richard; Groffman, Peter; Yesilonis, Ian; Hernandez, Luis (2002). "Soil carbon pools and fluxes in urban ecosystems" (PDF). Environmental Pollution. 116 (Supplement 1): S107–S118. doi:10.1016/S0269-7491(01)00263-9.
15. Davidson, Eric A.; Janssens, Ivan A. (2006). "Temperature sensitivity of soil carbon decomposition and feedbacks to climate change" (PDF). Nature. 440 (9 March 2006): 165□173. doi:10.1038/nature04514.
16. Powlson, David (2005). "Climatology: will soil amplify climate change?". Nature. 433 (20 January 2005): 204□205. doi:10.1038/433204a. (subscription required (help)).
17. Bradford, Mark A.; Wieder, William R.; Bonan, Gordon B.; Fierer, Noah; Raymond, Peter A.; Crowther, Thomas W. (2016). "Managing uncertainty in soil carbon feedbacks to climate change" (PDF). Nature Climate Change. 6 (27 July 2016): 751–758. doi:10.1038/nclimate3071.

18. Dominati, Estelle; Patterson, Murray; Mackay, Alec (2010). “A framework for classifying and quantifying the natural capital and ecosystem services of soils”(PDF). Ecological Economics. 69 (9): 1858–1868. doi:10.1016/j.ecolecon.2010.05.002.
19. Dykhuizen, Daniel E. (1998). “Santa Rosalia revisited: why are there so many species of bacteria?” (PDF). Antonie van Leeuwenhoek. 73 (1): 25–33. doi:10.1023/A:1000665216662.
20. Torsvik, Vigdis; Øvreås, Lise (2002). “Microbial diversity and function in soil: from genes to ecosystems” (PDF). Current Opinion in Microbiology. 5 (3): 240–245. doi:10.1016/S1369-5274(02)00324-7.
21. Raynaud, Xavier; Nunan, Naoise (2014). “Spatial ecology of bacteria at the microscale in soil” (PDF). PLOS ONE. 9 (1): e87217. doi:10.1371/journal.pone.0087217.
22. Whitman, William B.; Coleman, David C.; Wiebe, William J. (1998). “Prokaryotes: the unseen majority” (PDF). Proceedings of the National Academy of Sciences of the USA. 95 (12): 6578–6583. doi:10.1073/pnas.95.12.6578.
23. Schlesinger, William H.; Andrews, Jeffrey A. (2000). “Soil respiration and the global carbon cycle” (PDF). Biogeochemistry. 48 (1): 7–20. doi:10.1023/A:1006247623877.
24. Denmead, Owen Thomas; Shaw, Robert Harold (1962). “Availability of soil water to plants as affected by soil moisture content and meteorological conditions”(PDF). Agronomy Journal. 54 (5): 385–390. doi:10.2134/agronj1962.00021962005400050005x.
25. House, Christopher H.; Bergmann, Ben A.; Stomp, Anne-Marie; Frederick, Douglas J. (1999). “Combining constructed wetlands and aquatic and soil filters for reclamation and reuse of water” (PDF). Ecological Engineering. 12 (1–2): 27–38. doi:10.1016/S0925-8574(98)00052-4.
26. Van Bruggen, Ariena H.C.; Semenov, Alexander M. (2000). “In search of biological indicators for soil health and disease suppression” (PDF). Applied Soil Ecology. 15 (1): 13–24. doi:10.1016/S0929-1393(00)00068-8.
27. “A citizen’s guide to monitored natural attenuation” (PDF). Retrieved 27 November 2016.
28. Linn, Daniel Myron; Doran, John W. (1984). “Effect of water-filled pore space on carbon dioxide and nitrous oxide production in tilled and nontilled soils” (PDF). Soil Science Society of America Journal. 48 (6): 1267–1272. doi:10.2136/sssaj1984.03615995004800060013x.
29. Miller, Raymond W.; Donahue, Roy Luther (1990). Soils: an introduction to soils and plant growth. Upper Saddle River, New Jersey: Prentice Hall. ISBN 978-0138202262.
30. Bot, Alexandra; Benites, José (2005). The importance of soil organic matter: key to drought-resistant soil and sustained food and production (PDF). Rome, Italy: Food and Agriculture Organization of the United Nations. ISBN 92-5-105366-9. Retrieved 27 November 2016.

31. Barber, Stanley A. (1995). "Chemistry of soil-nutrient associations". In Barber, Stanley A. Soil nutrient bioavailability: a mechanistic approach (2nd ed.). New York, New York: John Wiley & Sons. pp. 9–48. ISBN 978-0-471-58747-7.
32. Möller, Detlev (2015). "Chemistry under environmental conditions". In Möller, Detlev. Chemistry for environmental scientists (1st ed.). Berlin, Germany: Walter de Gruyter. pp. 7–52. doi:10.1515/9783110410013-004. ISBN 978-3-11-041001-3.
33. "Arizona Master Gardener Manual". Cooperative Extension, College of Agriculture, University of Arizona. Retrieved 4 December 2016.
34. Vannier, Guy (1987). "The porosphere as an ecological medium emphasized in Professor Ghilarov's work on soil animal adaptations". Biology and Fertility of Soils. 3 (1): 39□44. doi:10.1007/BF00260577.
35. Torbert, H. Allen; Wood, Wes (1992). "Effect of soil compaction and water-filled pore space on soil microbial activity and N losses" (PDF). Communications in Soil Science and Plant Analysis. 23 (11): 1321□1331. doi:10.1080/00103629209368668.
36. Simonson 1957, p. 17.
37. Fookes, Peter G. (1997). Tropical residual soils. London, United Kingdom: Geological Society of London. ISBN 978-1897799383.
38. "Soil and water". Food and Agriculture Organization of the United Nations. Retrieved 11 December 2016.
39. "Sources. Negative Charge:". Jan.ucc.nau.edu. Retrieved 7 November 2012.
40. Donahue, Miller & Shickluna 1977, p. 4.
41. Kellogg 1957, p. 1.
42. Brady, Nyle (1984). The Nature and Properties of Soils (9th ed.). USA: Macmillan Publishing Co. pp. 4–7. ISBN 0023133406.
43. Kellogg 1957, p. 3.
44. Kellogg 1957, p. 2.
45. Kellogg 1957, p. 4.
46. "nitrogen fixation." Encyclopædia Britannica. Encyclopædia Britannica 2009 Deluxe Edition. Chicago: Encyclopædia Britannica, 2009.
47. Kellogg 1957, pp. 1–4.
48. Michael E. Ritter. Factors Affecting Soil Development, Soil Systems, The Physical Environment: an Introduction to Physical Geography, University of Wisconsin, Stevens Point, 1 October 2009, retrieved 3 January 2012.
49. "Soils". Natural Resource Conservation Service. United States Department of Agriculture. Retrieved 26 May 2013.
50. Van Schöll, Laura; Smits, Mark M.; Hoffland, Ellis (2006). "Ectomycorrhizal weathering of the soil minerals muscovite and hornblende". New Phytologist. 171 (4): 805–814. doi:10.1111/j.1469-8137.2006.01790.x. PMID 16918551.
51. Donahue, Miller & Shickluna 1977, pp. 20–21.
52. Donahue, Miller & Shickluna 1977, p. 21.

53. Donahue, Miller & Shickluna 1977, p. 24.
54. Gilluly, Waters, Woodford (1975). Principles of Geology (4th ed.). USA: W.H. Freeman. p. 215. ISBN 978-0716702696.
55. Donahue, Miller & Shickluna 1977, pp. 28–31.
56. Donahue, Miller & Shickluna 1977, pp. 31–33.
57. H.B. Milford, A.J.E. McGaw and K.J. Nixon, Soil Data Entry Handbook for the NSW Soil and Land Information System (SALIS), 3rd ed., New South Wales Department of Land and Water Conservation Resource Information Systems Group, Parramatta, 2001, pdf pp. 30–32.
58. Gove Hambidge, "Climate and Man—A Summary", in Erwin Raisz, U.S. Department of Agriculture, Climate And Man, Part One, Yearbook of Agriculture 1941, Repr. Honolulu: University Press of the Pacific, 2004, ISBN 978-1-4102-1538-3, pp. 1–66, p. 27.
59. Donahue, Miller & Shickluna 1977, p. 35.
60. Copley, Jon (25 August 2005). "Millions of bacterial species revealed underfoot". New Scientist. Retrieved 19 April 2010.
61. Amber Dance (2008). "Soil ecology: What lies beneath". Nature. 455 (7214): 724–5. doi:10.1038/455724a. PMID 18843336.
62. ^ Jump up to:a b Roesch LF, Fulthorpe RR, Riva A, Casella G, Hadwin AK, Kent AD, Daroub SH, Camargo FA, Farmerie WG, Triplett EW; Fulthorpe; Riva; Casella; Hadwin; Kent; Daroub; Camargo; Farmerie; Triplett (2007). "Pyrosequencing enumerates and contrasts soil microbial diversity". The ISME Journal. 1 (4): 283–90. doi:10.1038/ismej.2007.53. PMC 2970868 . PMID 18043639.
63. Gans J, Wolinsky M, Dunbar J.; Wolinsky; Dunbar (2005). "Computational improvements reveal great bacterial diversity and high metal toxicity in soil". Science. 309 (5739): 1387–90. Bibcode:2005Sci...309.1387G. doi:10.1126/science.1112665. PMID 16123304.
64. Simonson 1957, pp. 20, 21.
65. Donahue, Miller & Shickluna 1977, p. 26.
66. Nyle C. Brady & Ray R. Weil (2009). Elements of the Nature and Properties of Soils (3rd Edition). Prentice Hall. ISBN 9780135014332.
67. R. B. Brown (September 2003). "Soil Texture" (PDF). Fact Sheet SL-29. University of Florida, Institute of Food and Agricultural Sciences. Retrieved 8 July 2008.
68. Russell 1957, pp. 32–33.
69. Flemming 1957, p. 331.
70. Donahue, Miller & Shickluna 1977, p. 53.
71. Soil Survey Division Staff (1993). "Soil Structure". Handbook 18. Soil survey manual. Soil Conservation Service. U.S. Department of Agriculture. Archived from the original on 16 March 2008. Retrieved 8 July 2008.

72. Donahue, Miller & Shickluna 1977, pp. 55–56.
73. "Soil Structure - Physical Properties". The Cooperative Soil Survey. Retrieved 19 October 2012.
74. Donahue, Miller & Shickluna 1977, pp. 59–61.
75. Donahue, Miller & Shickluna 1977, p. 60.
76. "Physical aspects of crop productivity". www.fao.org. FAO. Retrieved 2 July 2016.
77. Donahue, Miller & Shickluna 1977, pp. 62–63.
78. "Soils - Part 2: Physical Properties of Soil and Soil Water". Retrieved 19 October 2012.
79. Johnson, T.A., T.R. Ellsworth, R.J.M. Hudson, and G.K. Sims. 2013. Diffusion Limitation for Atrazine Biodegradation in Soil. Advances in Microbiology. 3(5): 412-420.
80. Donahue, Miller & Shickluna 1977, pp. 62–63,565–567.
81. Taylor, H.M. 1983. Managing root systems for efficient water use: an overview, p. 87–113 in Taylor, H.H. et al. (Eds.). Limitations to Efficient Water Use in Crop Production. ASA, CCSA, and SSSA, Madison WI.
82. Gill, D. 1975. Influence of white spruce trees on permafrost-table microtopography, Mackenzie River Delta. Can. J. Earth Sci. 12(2):263–272.
83. Mityga, H.G.; Lanphear, F.O. 1971. Factors influencing the cold hardiness of Taxus cuspidata roots. J. Amer. Soc. Hort. Sci. 96:83–86.
84. Havis, J.R. 1976. Root hardiness of woody ornamentals. HortScience 11:385–386.
85. Lyr, H.; Hoffmann, G. 1967. Growth rates and growth periodicity of tree roots. p. 181–236 in Romberger, J.A.; Mikola, P. (Eds.). International Review of Forest Research, Vol. 2, Academic Press, New York NY.
86. Chalupa, V.; Fraser, D.A. 1968. Effect of soil and air temperature on soluble sugars and growth of white spruce (Picea glauca) seedlings. Can. J. Bot. 46:65–69.
87. Heninger, R.L.; White, D.P. 1974. Tree seedling growth at different soil temperatures. For. Sci. 20:363–367.
88. Ritchie, G.A.; Dunlap, J.R. 1980. Root growth potential: its development and expression in forest tree seedlings. New Zealand J. For. Sci. 10:218–248.
89. Binder, W.D.; Spittlehouse, D.L.; Draper, D.A. 1988. Post-planting studies of the physiology of white spruce 1984–1985, E.P. 966. B.C. Min. For., Res. Branch, Victoria BC, Progr. Rep. 5, unpubl. 85 p. [Coates et al. 1994 give date as 1987, Binder bibliog at R986 gives 1988]
90. Landhäusser, S.M.; DesRochers, A.; and Lieffers, V.J. 2001. A comparison of growth and physiology in Picea glauca and Populus tremuloides at different soil temperatures. Can. J. For. Res. 31:1922–1929.
91. Tryon, P.R.; Chapin, F.S. 1983. Temperature control over root growth and root biomass in taiga forest trees. Can. J. For. Res. 13(5):827–833.

92. Landhäusser, S.M.; Silins, U.; Lieffers, V.J.; Liu, W. 2003. Response of Populus tremuloides, Populus balsamifera, Betula papyrifera and Picea glauca seedlings to low soil temperature and water-logged soil conditions. Scan. J. For. Res. 18:391–400. (Cited Green 2004)
93. Turner, N.C.; Jarvis, P.G. 1975. Photosynthesis in Sitka spruce (Picea sitchensis Bong.). IV. Response to soil temperature. J. Appl. Ecol. 12:561–576.
94. Day, T.A.; DeLucia, E.H.; Smith, W.K. 1990. Effect of soil temperature on stem flow, shoot gas exchange and water potential of Picea engelmannii (Parry) during snowmelt. Oecologia 84(4):474–481.
95. Green, D.S. 2004. Describing condition-specific determinants of competition in boreal and sub-boreal mixedwood stands. For. Chron. 80(6):736–742.
96. Donahue, Miller & Shickluna 1977, p. 71.
97. "Arizona Master Gardener Manual". Cooperative Extension, College of Agriculture, University of Arizona. p. Chapter 2, pp 4–8. Retrieved 27 May 2013.
98. "The Color of Soil". United States Department of Agriculture - Natural Resources Conservation Service. Archived from the original on 16 March 2008. Retrieved 8 July 2008.
99. "Electrical Design, Cathodic Protection". United States Army Corps of Engineers. 22 April 1985. Archived from the original on 12 June 2008. Retrieved 2 July 2008.
100. "The why and how to testing the Electrical Conductivity of Soils | Resources". Retrieved 19 December 2010.
101. R. J. Edwards (15 February 1998). "Typical Soil Characteristics of Various Terrains". Retrieved 2 July 2008.
102. Donahue, Miller & Shickluna 1977, p. 72.
103. Wadleigh 1957, p. 48.
104. Richards & Richards 1957, p. 50.
105. Richards & Richards 1957, p. 56.
106. Wadleigh 1957, p. 39.
107. Richards & Richards 1957, p. 52.
108. Donahue, Miller & Shickluna 1977, pp. 72-74.
109. "CHAPTER 2 - SOIL AND WATER". Fao.org. Retrieved 7 November 2012.
110. Donahue, Miller & Shickluna 1977, pp. 72–75.
111. Wadleigh 1957, p. 47.
112. Donahue, Miller & Shickluna 1977, pp. 75–76.
113. Donahue, Miller & Shickluna 1977, pp. 76–77.
114. Donahue, Miller & Shickluna 1977, p. 80.
115. Donahue, Miller & Shickluna 1977, p. 85.
116. Donahue, Miller & Shickluna 1977, p. 86.

117. Donahue, Miller & Shickluna 1977, p. 88.
118. Brehm, Denise (11 December 2008). "CEE researchers explain mystery of gravity fingers". MIT Department of Civil & Environmental Engineering. MIT. Retrieved 31 October 2012.
119. "Urban Trees Enhance Water Infiltration". Fisher, Madeline. The American Society of Agronomy. 17 November 2008. Retrieved 31 October 2012.
120. "Major floods recharge aquifers". University of New South Wales Science. 24 January 2011. Retrieved 31 October 2012.
121. Donahue, Miller & Shickluna 1977, p. 90.
122. Donahue, Miller & Shickluna 1977, p. 91.
123. Donahue, Miller & Shickluna 1977, p. 92.
124. Wadleigh 1957, p. 46.
125. Donahue, Miller & Shickluna 1977, p. 94.
126. Donahue, Miller & Shickluna 1977, pp. 97–99.
127. Russell 1957, pp. 35–36.
128. http://atmo.tamu.edu/class/geos489/lecture6/soiltracegasrev.pdf
129. College of Tropical Agriculture and Human Resources. "Soil Mineralogy". www.ctahr.hawaii.edu. University of Hawai‘i. Retrieved 7 August 2014.
130. ^ Jump up to:a b Russell, E. Walter (1973). Soil conditions and plant growth. (10th ed.). London: Longman. pp. 67–70. ISBN 0582440483.
131. Donahue, Miller & Shickluna 1977, pp. 101–102.
132. Simonson 1957, p. 19.
133. Donahue, Miller & Shickluna 1977, p. 102.
134. Russell 1957, p. 33.
135. Donahue, Miller & Shickluna 1977, pp. 102–107.
136. Donahue, Miller & Shickluna 1977, pp. 101–107.
137. Donahue, Miller & Shickluna 1977, p. 107.
138. Donahue, Miller & Shickluna 1977, p. 108.
139. Russell 1957, pp. 33, 34.
140. Coleman & Mehlich 1957, p. 74.
141. Donahue, Miller & Shickluna 1977, pp. 108–110.
142. Dean 1957, p. 82.
143. Allison 1957, p. 90.
144. Reitemeier 1957, p. 103.
145. Donahue, Miller & Shickluna 1977, p. 110.
146. Coleman & Mehlich 1957, p. 73.
147. Holmes & Brown 1957, p. 112.

148. Donahue, Miller & Shickluna 1977, p. 111.
149. Olsen & Fried 1957, p. 96.
150. Reitemeier 1957, p. 101.
151. Donahue, Miller & Shickluna 1977, pp. 103–112.
152. Simonson 1957, pp. 18, 21–24, 29.
153. Russell 1957, pp. 32, 35.
154. Donahue, Miller & Shickluna 1977, p. 112.
155. Russell 1957, p. 35.
156. Allaway 1957, p. 69.
157. Lehmann, J. "Terra Preta de Indio". University of Cornell, Dept. of Crop and Soil Sciences. Retrieved 30 March 2013.
158. Donahue, Miller & Shickluna 1977, p. 103–106.
159. Donahue, Miller & Shickluna 1977, p. 114.
160. Donahue, Miller & Shickluna 1977, pp. 115–116.
161. Chang, Raymond (1984). Chemistry. Random House, Inc. p. 424. ISBN 0-394-32983-X.
162. Donahue, Miller & Shickluna 1977, pp. 116–117.
163. Donahue, Miller & Shickluna 1977, pp. 116–119.
164. Donahue, Miller & Shickluna 1977, pp. 119–120.
165. Donahue, Miller & Shickluna 1977, pp. 120–121.
166. Dean 1957, p. 80.
167. Russel 1957, pp. 123–125.
168. Dean 1957, pp. 80, 81.
169. Roy, R.N.; Finck, A.; Blair, G.J.; Tandon, H.L.S. (2006). "Chapter 4: Soil fertility and crop production". Plant nutrition for food security: a guide for integrated nutrient management (PDF). Rome: Food and Agriculture Organization of the United Nations. pp. 43–90. ISBN 92-5-105490-8. Retrieved 20 June 2016.
170. Donahue, Miller & Shickluna 1977, pp. 123–131.
171. Donahue, Miller & Shickluna 1977, p. 125.
172. Donahue, Miller & Shickluna 1977, p. 126.
173. "Lecture 22". Northern Arizona University. Retrieved 22 March 2013.
174. Donahue, Miller & Shickluna 1977, pp. 123–128.
175. Wadleigh 1957, p. 41.
176. Broadbent 1957, p. 153.
177. Donahue, Miller & Shickluna 1977, p. 128.
178. Allison 1957, pp. 85–94.
179. Broadbent 1957, pp. 152–155.

180. Donahue, Miller & Shickluna 1977, pp. 128–131.
181. Donahue, Miller & Shickluna 1977, pp. 129–130.
182. Donahue, Miller & Shickluna 1977, p. 145.
183. Donahue, Miller & Shickluna 1977, pp. 128–129.
184. Allison 1957, p. 87.
185. Donahue, Miller & Shickluna 1977, p. 130.
186. Donahue, Miller & Shickluna 1977, p. 131.
187. Donahue, Miller & Shickluna 1977, pp. 134–135.
188. Reitemeier 1957, pp. 101–104.
189. Donahue, Miller & Shickluna 1977, pp. 135–136.
190. Donahue, Miller & Shickluna 1977, p. 136.
191. Jordan & Reisenauer 1957, pp. 107.
192. Holmes & Brown 1957, pp. 111.
193. Sherman 1957, pp. 135.
194. Seatz & Jurinak 1957, pp. 115.
195. Reuther 1957, pp. 128.
196. Russel 1957, pp. 121.
197. Stout & Johnson 1957, pp. 146.
198. Stout & Johnson 1957, pp. 141.
199. Donahue, Miller & Shickluna 1977, pp. 136–137.
200. Stout & Johnson 1957, pp. 107.
201. Pimentel, D.; et al. (1995). "Environmental and economic costs of soil erosion and conservation benefits". Science. 267 (24): 1117–22. Bibcode:1995Sci...267.1117P. doi:10.1126/science.267.5201.1117. PMID 17789193.
202. Foth, Henry D. (1984). Fundamentals of soil science. New York: Wiley. p. 151. ISBN 0-471-88926-1.
203. Gilluly, Waters, Woodford (1975). Principles of Geology (4th ed.). USA: W.H. Freeman. p. 216. ISBN 978-0716702696.
204. Verkaik, Eric; Jongkind, Anne G.; Berendse, Frank (2006). "Short-term and long-term effects of tannins on nitrogen mineralization and litter decomposition in kauri (Agathis australis (D. Don) Lindl.) forests". Plant and Soil. 287: 337. doi:10.1007/s11104-006-9081-8.
205. Fierer, N.; Schimel, Joshua P.; Cates, Rex G.; Zou, Jiping (2001). "Influence of balsam poplar tannin fractions on carbon and nitrogen dynamics in Alaskan taiga floodplain soils". Soil Biology and Biochemistry. 33 (12–13): 1827. doi:10.1016/S0038-0717(01)00111-0.
206. Wagai, Rota; Mayer, Lawrence M.; Kitayama, Kanehiro; Knicker, Heike (2008). "Climate and parent material controls on organic matter storage in surface soils: A three-pool, density-separation approach". Geoderma. 147: 23–33. doi:10.1016/j.

geoderma.2008.07.010.

207. Minayeva, T. Yu.; Trofimov, S. Ya.; Chichagova, O.A.; Dorofeyeva, E.I.; Sirin, A.A.; Glushkov, I.V.; Mikhailov, N.D.; Kromer, B. (2008). "Carbon accumulation in soils of forest and bog ecosystems of southern Valdai in the Holocene". Biology Bulletin. 35 (5): 524–532. doi:10.1134/S1062359008050142.

208. Sanchez, Pedro A. (1976). Properties and management of soils in the tropics. New York: Wiley. ISBN 0471752002.

209. Paul, E.A. (1997). Soil organic matter in temperate agroecosystems : long-term experiments in North America. Boca Raton: CRC Press. p. 80. ISBN 9780849328022.

210. Retallack, G. J. (1990). Soils of the past : an introduction to paleopedology. Boston: Unwin Hyman. p. 32. ISBN 978-0-04-445757-2.

211. Buol, S.W. (1990). Soil genesis and classification. Ames, Iowa: Iowa State University Press. p. 36. doi:10.1081/E-ESS. ISBN 0-8138-2873-2.

212. IUSS Working Group WRB (2014). World Reference Base for Soil Resources 2014. International soil classification system for naming soils and creating legends for soil maps (PDF) (3rd ed.). Rome: FAO. ISBN 978-92-5-108370-3. Retrieved 29 August 2014.

213. bhttp://eusoils.jrc.ec.europa.eu/esdb_archive/eusoils_docs/other/EUR23439.pdfSoils of the European Union by the EU Institute for Environment and Sustainability. Accessed on 8 October 2013

214. The Soil Orders, Department of Environmental Sciences, University of Virginia, retrieved 23 October 2012.

215. Donahue, Miller & Shickluna 1977, pp. 411–432.

216. Soil Survey Staff (1999). Soil taxonomy: A Basic System of Soil Classification for Making and Interpreting Soil Surveys. 2nd edition. Natural Resources Conservation Service. U.S. Department of Agriculture Handbook 436. (PDF). USDA: United States Dept. of Agriculture, Naturel Resources Conservation Service. Retrieved 10 March 2013.

217. The Twelve Soil Orders: Soil Taxonomy, Soil & Land Resources Division, College of Agricultural and Life Sciences, University of Idaho

218. Donahue, Miller & Shickluna 1977, p. 409.

219. Donahue, Miller & Shickluna 1977, pp. 409-410.

220. Donahue, Miller & Shickluna 1977, p. 410.

221. Leake, Simon; Haege, Elke (2014). Soils for Landscape Development. CSIRO Publishing. ISBN 9780643109643.

222. Ponge, Jean-François (2003). "Humus forms in terrestrial ecosystems: a framework to biodiversity" (PDF). Soil Biology and Biochemistry. 35 (7): 935–945. doi:10.1016/S0038-0717(03)00149-4.

223. De Deyn, Gerlinde B.; Van der Putten, Wim H. (2005). "Linking aboveground and belowground diversity". Trends in Ecology & Evolution. 20 (11): 625–633.

doi:10.1016/j.tree.2005.08.009. PMID 16701446.

224. Hansen, J.; et al. (2008). "Target atmospheric CO2: Where should humanity aim?". Open Atmospheric Science Journal. 2: 217–31. arXiv:0804.1126 . Bibcode:2008OASJ....2..217H. doi:10.2174/1874282300802010217

225. Lal, R. (11 June 2004). "Soil Carbon Sequestration Impacts on Global Climate Change and Food Security". Science. 304 (5677): 1623–1627. Bibcode:2004Sci...304.1623L. doi:10.1126/science.1097396. PMID 15192216.

226. Blakeslee, Thomas R. (24 February 2010). "Greening Deserts for Carbon Credits". Renewable Energy World.com. Retrieved 23 October 2012.

227. Setz, EZF; Enzweiler J; Solferini VN; Amendola MP; Berton RS (1999). "Geophagy in the golden-faced saki monkey (Pithecia pithecia chrysocephala) in the Central Amazon". Journal of Zoology. 247 (1): 91–103. doi:10.1111/j.1469-7998.1999.tb00196.x. (subscription required (help)).

228. Kohne, John Maximilian; Koehne, Sigrid; Simunek, Jirka (2009). "A review of model applications for structured soils: a) Water flow and tracer transport". Journal of Contaminant Hydrology. 104 (1–4): 4–35. Bibcode:2009JCHyd.104....4K. doi:10.1016/j.jconhyd.2008.10.002. PMID 19012994.

229. Diplock, EE; Mardlin DP; Killham KS; Paton GI (2009). "Predicting bioremediation of hydrocarbons: laboratory to field scale". Environmental Pollution. 157 (6): 1831–1840. doi:10.1016/j.envpol.2009.01.022. PMID 19232804.

230. Moeckel, Claudia; Nizzetto, Luca; Di Guardo, Antonio; Steinnes, Eiliv; Freppaz, Michele; Filippa, Gianluca; Camporini, Paolo; Benner, Jessica; Jones, Kevin C. (2008). "Persistent organic pollutants in boreal and montane soil profiles: distribution, evidence of processes and implications for global cycling". Environmental Science and Technology. 42 (22): 8374–8380. Bibcode:2008EnST...42.8374M. doi:10.1021/es801703k. PMID 19068820.

231. Rezaei, Khalil; Guest, Bernard; Friedrich, Anke; Fayazi, Farajollah; Nakhaei, Mohamad; Aghda, Seyed Mahmoud Fatemi; Beitollahi, Ali (2009). "Soil and sediment quality and composition as factors in the distribution of damage at the December 26, 2003, Bam area earthquake in SE Iran (M (s)=6.6)". Journal of Soils and Sediments. 9: 23–32. doi:10.1007/s11368-008-0046-9.

232. Johnson, D.L.; Ambrose, S.H.; Bassett, T.J.; Bowen, M.L.; Crummey, D.E.; Isaacson, J.S.; Johnson, D.N.; Lamb, P.; Saul, M.; Winter-Nelson, A. E. (1997). "Meanings of environmental terms". Journal of Environmental Quality. 26 (3): 581–589. doi:10.2134/jeq1997.00472425002600030002x.

233. Jones, j. a. a. (1976). "Soil piping and stream channel initiation". Water Resources Research. 7 (3): 602–610. Bibcode:1971WRR.....7..602J. doi:10.1029/WR007i003p00602.

234. Dooley, Alan (June 2006). "Sandboils 101: Corps has experience dealing with common flood danger". Engineer Update. US Army Corps of Engineers. Archived from the original on 18 April 2008. Retrieved 14 May 2008.

235. ILRI (1989). "Effectiveness and Social/Environmental Impacts of Irrigation Projects: a Review" (pdf). In: Annual Report 1988 of the International Institute

for Land Reclamation and Improvement (ILRI). Wageningen, The Netherlands. pp. 18–34.

236. Drainage Manual: A Guide to Integrating Plant, Soil, and Water Relationships for Drainage of Irrigated Lands. Interior Dept., Bureau of Reclamation. 1993. ISBN 0-16-061623-9.

237. "Free articles and software on drainage of waterlogged land and soil salinity control". Retrieved 28 July 2010.

238. "Improving soils and boosting yields in Thailand" (PDF). Success Stories. International Water Management Institute (2). 2010. doi:10.5337/2011.0031.

239. "Provide for your garden's basic needs ... and the plants will take it from there". USA Weekend. 10 March 2011.

CHAPTER 2

Soil Physics and Morphology

Chapter Contents

- **Physical Properties of Soils**
- **Consumptive Use and Water Efficiency**
- **Micromorphology**
- **Soil Mechanics**
- **Genesis and Composition of Soils**
- **Effective Stress and Capillarity: Hydrostatic Conditions**
- **Soil Classification**
- **Seepage: Steady State Flow of Water**
- **Consolidation: Transient Flow of Water**
- **Shear Behavior: Stiffness and Strength**
- **Applications**
- **Recent Developments**

Soils are an integral part of ecological and agricultural systems. Soils occur as a continuum across the landscape and their properties change in a predictable fashion both across the landscape and with depth. An understanding of soil morphology and the spatial distribution of soil properties is critical in the land-use decision making process. This two week course will cover topics such as soil morphology, soil forming factors, soil genesis, soil classification, soil mapping, and soil geomorphology.

This course is taught via distance learning, but the instructor will supplement lecture material with additional readings and practical examples to illustrate the concepts and provide practical examples of how the concepts are used in practice. This course is designed to complement the students existing knowledge of soil science and help the student understand the principles behind the Soil Science Performance Objectives, which define the practice of soil science.

PHYSICAL PROPERTIES OF SOILS

The physical properties of soils, in order of decreasing importance, are texture, structure, density, porosity, consistency, temperature, colour and resistivity. Soil texture is determined by the relative proportion of the three kinds of soil mineral particles, called soil separates:

sand, silt, and clay. At the next larger scale, soil structures called peds are created from the soil separates when iron oxides, carbonates, clay, silica and humus, coat particles and cause them to adhere into larger, relatively stable secondary structures. Soil density, particularly bulk density, is a measure of soil compaction. Soil porosity consists of the void part of the soil volume and is occupied by gases or water. Soil consistency is the ability of soil to stick together. Soil temperature and colour are self-defining. Resistivity refers to the resistance to conduction of electric currents and affects the rate of corrosion of metal and concrete structures. These properties may vary through the depth of a soil profile. Most of these properties determine the aeration of the soil and the ability of water to infiltrate and to be held within the soil.

Influence of Soil Texture Separates on Some Properties of Soils

Property/behavior	**Sand**	**Silt**	**Clay**
Water-holding capacity	Low	Medium to high	High
Aeration	Good	Medium	Poor
Drainage rate	High	Slow to medium	Very slow
Soil organic matter level	Low	Medium to high	High to medium
Decomposition of organic matter	Rapid	Medium	Slow
Warm-up in spring	Rapid	Moderate	Slow
Compactability	Low	Medium	High
Susceptibility to wind erosion	Moderate (High if fine sand)	High	Low
Susceptibility to water erosion	Low (unless fine sand)	High	Low if aggregated, otherwise high
Shrink/Swell Potential	Very Low	Low	Moderate to very high
Sealing of ponds, dams, and landfills	Poor	Poor	Good
Suitability for tillage after rain	Good	Medium	Poor
Pollutant leaching potential	High	Medium	Low (unless cracked)
Ability to store plant nutrients	Poor	Medium to High	High

Resistance to pH change	Low	Medium	High

Texture

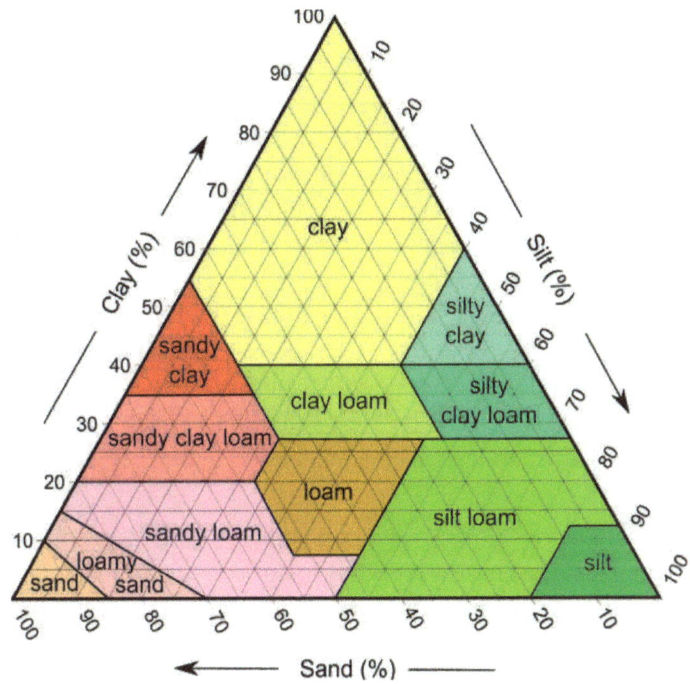

Soil types by clay, silt and sand composition as used by the USDA

Iron-rich soil near Paint Pots in Kootenay National Park, Canada

The mineral components of soil are sand, silt and clay, and their relative proportions determine a soil's texture. Properties that are influenced by soil texture, include porosity, permeability, infiltration, shrink-swell rate, water-holding capacity, and susceptibility to erosion. In the illustrated USDA textural classification triangle, the only soil in which neither sand, silt nor clay predominates is called "loam". While even pure sand, silt or clay may be considered a soil, from the perspective of food production a loam soil with a small amount of organic material is considered ideal. The mineral constituents of a loam soil might be 40% sand, 40% silt and the balance 20% clay by weight. Soil texture affects soil behaviour, in particular its retention capacity for nutrients and water.

Sand and silt are the products of physical and chemical weathering of the parent rock; clay, on the other hand, is a most often the product of the precipitation of the dissolved parent rock as a secondary mineral. It is the surface area to volume ratio (specific surface area) of soil particles and the unbalanced ionic charges within those that determine their role in the fertility of soil, as measured by its cation exchange capacity. Sand is least active, followed by silt; clay is the most active. Sand's greatest benefit to soil is that it resists compaction and increases a soil's porosity. Silt is mineralogically like sand but with its higher specific surface area it is more chemically active than sand. But it is the clay content of soil, with its very high specific surface area and generally large number of negative charges, that gives a soil its high retention capacity for water and nutrients. Clay soils also resist wind and water erosion better than silty and sandy soils, as the particles bond tightly to each other.

Sand is the most stable of the mineral components of soil; it consists of rock fragments, primarily quartz particles, ranging in size from 2.0 to 0.05 mm (0.0787 to 0.0020 in) in diameter. Silt ranges in size from 0.05 to 0.002 mm (0.002 to 0.00008 in). Clay cannot be resolved by optical microscopes as its particles are 0.002 mm (7.9×10^{-5} in) or less in diameter and a thickness of only 10 angstroms (10^{-10} m). In medium-textured soils, clay is often washed downward through the soil profile and accumulates in the subsoil.

Soil components larger than 2.0 mm (0.079 in) are classed as rock and gravel and are removed before determining the percentages of the remaining components and the texture class of the soil, but are included in the name. For example, a sandy loam soil with 20% gravel would be called gravelly sandy loam.

When the organic component of a soil is substantial, the soil is called organic soil rather than mineral soil. A soil is called organic if:

- Mineral fraction is 0% clay and organic matter is 20% or more
- Mineral fraction is 0% to 50% clay and organic matter is between 20% and 30%
- Mineral fraction is 50% or more clay and organic matter 30% or more.

Structure

The clumping of the soil textural components of sand, silt and clay causes aggregates to form and the further association of those aggregates into larger units creates soil structures called peds (a contraction of the word pedolith). The adhesion of the soil textural components by organic substances, iron oxides, carbonates, clays, and silica, and the breakage of those aggregates from expansion-contraction, caused by freezing-thawing and wetting-drying cycles, shape soil into distinct geometric forms. The peds evolve into units which have various shapes, sizes and degrees of development. A soil clod, however, is not a ped but rather a mass of soil that results from mechanical disturbance of the soil. Soil structure affects aeration, water movement, conduction of heat, plant root growth and resistance to erosion. Water, in turn, has its strongest effect on soil structure due to its solution and precipitation of minerals and its effect on plant growth.

Soil structure often gives clues to its texture, organic matter content, biological activity, past soil evolution, human use, and the chemical and mineralogical conditions under which the soil formed. While texture is defined by the mineral component of a soil and is an innate property of the soil that does not change with agricultural activities, soil structure can be improved or destroyed by the choice and timing of farming practices.

Soil structural classes:

1. Types: Shape and arrangement of peds
    1. Platy: Peds are flattened one atop the other 1–10 mm thick. Found in the A-horizon of forest soils and lake sedimentation.
    2. Prismatic and Columnar: Prismlike peds are long in the vertical dimension, 10–100 mm wide. Prismatic peds have flat tops, columnar peds have rounded tops. Tend to form in the B-horizon in high sodium soil where clay has accumulated.
    3. Angular and subangular: Blocky peds are imperfect cubes, 5–50 mm, angular have sharp edges, subangular have rounded edges. Tend to form in the B-horizon where clay has accumulated and indicate poor water penetration.
    4. Granular and Crumb: Spheroid peds of polyhedrons, 1–10 mm, often found in the A-horizon in the presence of organic material. Crumb peds are more porous and are considered ideal.
2. Classes: Size of peds whose ranges depend upon the above type
    1. Very fine or very thin: <1 mm platy and spherical; <5 mm blocky; <10 mm prismlike.
    2. Fine or thin: 1–2 mm platy, and spherical; 5–10 mm blocky; 10–20 mm prismlike.
    3. Medium: 2–5 mm platy, granular; 10–20 mm blocky; 20-50 prismlike.
    4. Coarse or thick: 5–10 mm platy, granular; 20–50 mm blocky; 50–100 mm prismlike.
    5. Very coarse or very thick: >10 mm platy, granular; >50 mm blocky; >100 mm prismlike.
3. Grades: Is a measure of the degree of development or cementation within the peds that results in their strength and stability.

1. Weak: Weak cementation allows peds to fall apart into the three textural constituents, sand, silt and clay.

2. Moderate: Peds are not distinct in undisturbed soil but when removed they break into aggregates, some broken aggregates and little unaggregated material. This is considered ideal.

3. Strong:Peds are distinct before removed from the profile and do not break apart easily.

4. Structureless: Soil is entirely cemented together in one great mass such as slabs of clay or no cementation at all such as with sand.

At the largest scale, the forces that shape a soil›s structure result from swelling and shrinkage that initially tend to act horizontally, causing vertically oriented prismatic peds. Clayey soil, due to its differential drying rate with respect to the surface, will induce horizontal cracks, reducing columns to blocky peds. Roots, rodents, worms, and freezing-thawing cycles further break the peds into a spherical shape.

At a smaller scale, plant roots extend into voids and remove water causing the open spaces to increase, thereby decreasing aggregate size. At the same time, roots, fungal hyphae, and earthworms create microscopic tunnels that break up peds.

At an even smaller scale, soil aggregation continues as bacteria and fungi exude sticky polysaccharides which bind soil into smaller peds. The addition of the raw organic matter that bacteria and fungi feed upon encourages the formation of this desirable soil structure.

At the lowest scale, the soil chemistry affects the aggregation or dispersal of soil particles. The clay particles contain polyvalent cations which give the faces of clay layers localized negative charges. At the same time, the edges of the clay plates have a slight positive charge, thereby allowing the edges to adhere to the negative charges on the faces of other clay particles or to flocculate (form clumps). On the other hand, when monovalent ions, such as sodium, invade and displace the polyvalent cations, they weaken the positive charges on the edges, while the negative surface charges are relatively strengthened. This leaves negative charge on the clay faces that repel other clay, causing the particles to push apart, and by doing,the flocculation of clay particles into larger, open assemblages. As a result, the clay disperses and settles into voids between peds, causing those to close. In this way the open structure of the soil is destroyed and the soil is made impenetrable to air and water. Such sodic soil tends to form columnar structures near the surface.

Density

Soil particle density is typically 2.60 to 2.75 grams per cm^3 and is usually unchanging for a given soil. Soil particle density is lower for soils with high organic matter content, and is higher for soils with high iron-oxides content. Soil bulk density is equal to the dry mass of the soil divided by the volume of the soil; i.e., it includes air space and organic materials of the soil volume. The soil bulk density of cultivated loam is about 1.1 to 1.4 g/cm^3 (for comparison water is 1.0 g/cm^3). Soil bulk density is highly variable for a given soil. A lower bulk density by itself does not indicate suitability for plant growth due to the influence of soil texture and structure. A high bulk density is indicative of either soil compaction or high sand content. Soil bulk density is inherently always less than the soil particle density.

Representative bulk densities of soils. The percentage pore space was calculated using 2.7 g/cm3 for particle density except for the peat soil, which is estimated.

Soil treatment and identification	Bulk density g/cm^3	Pore space %
Tilled surface soil of a cotton field	1.3	51
Trafficked inter-rows where wheels passed surface	1.67	37
Traffic pan at 25 cm deep	1.7	36
Undisturbed soil below traffic pan, clay loam	1.5	43
Rocky silt loam soil under aspen forest	1.62	40
Loamy sand surface soil	1.5	43
Decomposed peat	0.55	65

Porosity

Pore space is that part of the bulk volume of soil that is not occupied by either mineral or organic matter but is open space occupied by either gases or water. In a productive, medium-textured soil the total pore space is typically about 50% of the soil volume. Pore size varies considerably; the smallest pores (cryptopores; <0.1 μm) hold water too tightly for use by plant roots; plant-available water is held in ultramicropores, micropores and mesopores (0.1-75 μm); and macropores (>75 μm) are generally air-filled when the soil is at field capacity.

Soil texture determines total volume of the smallest pores; clay soils have smaller pores, but more total pore space than sands. Soil structure has a strong influence on the larger pores that affect soil aeration, water infiltration and drainage. Tillage has the short-term benefit of temporarily increasing the number of pores of largest size, but these can be rapidly degraded by the destruction of soil aggregation.

The pore size distribution affects the ability of plants and other organisms to access water and oxygen; large, continuous pores allow rapid transmission of air, water and dissolved nutrients through soil, and small pores store water between rainfall or irrigation events. Pore size variation also compartmentalizes the soil pore space such that many microorganisms are not in direct competition with one another, which may explain not only the large number of species present, but the fact that functionally redundant microorganisms (organisms with the same ecological niche) can co-exist within the same soil.

Consistency

Consistency is the ability of soil to stick to itself or to other objects (cohesion and adhesion respectively) and its ability to resist deformation and rupture. It is of approximate use in predicting cultivation problems and the engineering of foundations. Consistency is measured at three moisture conditions: air-dry, moist, and wet. In those conditions the consistency quality depends upon the clay content. In the wet state, the two qualities of stickiness and plasticity are assessed. A soil's resistance to fragmentation and crumbling is assessed in the dry state by rubbing the sample. Its resistance to shearing forces is assessed in the moist state by thumb and finger pressure. Additionally, the cemented consistency depends on cementation by substances other than clay, such as calcium carbonate, silica, oxides and salts; moisture content has little effect on its assessment. The measures of consistency border on subjective compared to other measures such as pH, since they employ the apparent feel of the soil in those states.

The terms used to describe the soil consistency in three moisture states and a last not affected by the amount of moisture are as follows:

- Consistency of Dry Soil: loose, soft, slightly hard, hard, very hard, extremely hard
- Consistency of Moist Soil: loose, very friable, friable, firm, very firm, extremely firm
- Consistency of Wet Soil: nonsticky, slightly sticky, sticky, very sticky; nonplastic, slightly plastic, plastic, very plastic
- Consistency of Cemented Soil: weakly cemented, strongly cemented, indurated (requires hammer blows to break up)

Soil consistency is useful in estimating the ability of soil to support buildings and roads. More precise measures of soil strength are often made prior to construction.

Temperature

Soil temperature depends on the ratio of the energy absorbed to that lost. Soil has a temperature range between -20 to 60 °C. Soil temperature regulates seed germination, plant and root growth and the availability of nutrients. Below 50 cm (20 in), soil temperature seldom changes and can be approximated by adding 1.8 °C (2 °F) to the mean annual air temperature. Soil temperature has important seasonal, monthly and daily variations. Fluctuations in soil temperature are much lower with increasing soil depth. Heavy mulching (a type of soil cover) can slow the warming of soil, and, at the same time, reduce fluctuations in surface temperature.

Most often, agricultural activities must adapt to soil temperatures by:

- maximizing germination and growth by timing of planting
- optimizing use of anhydrous ammonia by applying to soil below 10 °C (50 °F)
- preventing heaving and thawing due to frosts from damaging shallow-rooted crops
- preventing damage to desirable soil structure by freezing of saturated soils

- improving uptake of phosphorus by plants

Soil temperatures can be raised by drying soils or the use of clear plastic mulches. Organic mulches slow the warming of the soil.

There are various factors that affect soil temperature, such as water content, soil color, and relief (slope, orientation, and elevation), and soil cover (shading and insulation). The color of the ground cover and its insulating properties have a strong influence on soil temperature. Whiter soil tends to have a higher albedo than blacker soil cover, which encourages whiter soils to have lower soil temperatures. The specific heat of soil is the energy required to raise the temperature of soil by 1 °C. The specific heat of soil increases as water content increases, since the heat capacity of water is greater than that of dry soil. The specific heat of pure water is ~ 1 calorie per gram, the specific heat of dry soil is ~ 0.2 calories per gram, hence, the specific heat of wet soil is ~ 0.2 to 1 calories per gram. Also, a tremendous energy (~540 cal/g) is required to evaporate water (known as the heat of vaporization). As such, wet soil usually warms more slowly than dry soil - wet surface soil is typically 3 to 6 °C colder than dry surface soil.

Soil heat flux refers to the rate at which heat energy moves through the soil in response to a temperature difference between two points in the soil. The heat flux density is the amount of energy that flows through soil per unit area per unit time and has both magnitude and direction. For the simple case of conduction into or out of the soil in the vertical direction, which is most often applicable the heat flux density is:

$$q_x = -k\frac{\delta T}{\delta x}$$

In SI units

q is the heat flux density, in SI the units are $W{\cdot}m^{-2}$

k is the soils' conductivity, $W{\cdot}m^{-1}{\cdot}K^{-1}$. The thermal conductivity is sometimes a constant, otherwise an average value of conductivity for the soil condition between the surface and the point at depth is used.

δT is the temperature difference (temperature gradient) between the two points in the soil between which the heat flux density is to be calculated. In SI the units are kelvin, K.

δx is the distance between the two points within the soil, at which the temperatures are measured and between which the heat flux density is being calculated. In SI the units are meters m, and where x is measured positive downward.

Heat flux is in the direction opposite the temperature gradient, hence the minus sign. That is to say, if the temperature of the surface is higher than at depth x the negative sign will result in a positive value for the heat flux q, and which is interpreted as the heat being conducted into the soil.

Component	Thermal Conductivity (W·m-1·K-1)
Quartz	8.8
Clay	2.9
Organic matter	0.25
Water	0.57

Ice	2.4
Air	0.025
Dry soil	0.2-0.4
Wet soil	1-3

Soil temperature is important for the survival and early growth of seedlings. Soil temperatures affect the anatomical and morphological character of root systems.All physical, chemical, and biological processes in soil and roots are affected in particular because of the increased viscosities of water and protoplasm at low temperatures. In general, climates that do not preclude survival and growth of white spruce above ground are sufficiently benign to provide soil temperatures able to maintain white spruce root systems. In some northwestern parts of the range, white spruce occurs on permafrost sites and although young unlignified roots of conifers may have little resistance to freezing, less than half of the "secondary mature" root system of white spruce was killed by exposure to a temperature of 23.3 °C in multiple year experiment with containerized trees from local nurseries in Massachusetts.

Optimum temperatures for tree root growth range between 10 °C and 25 °C in general and for spruce in particular. In 2-week-old white spruce seedlings that were then grown for 6 weeks in soil at temperatures of 15 °C, 19 °C, 23 °C, 27 °C, and 31 °C; shoot height, shoot dry weight, stem diameter, root penetration, root volume, and root dry weight all reached maxima at 19 °C.

However, whereas strong positive relationships between soil temperature (5 °C to 25 °C) and growth have been found in trembling aspen and balsam poplar, white and other spruce species have shown little or no changes in growth with increasing soil temperature. Such insensitivity to soil low temperature may be common among a number of western and boreal conifers.

Color

Soil colour is often the first impression one has when viewing soil. Striking colours and contrasting patterns are especially noticeable. The Red River of the South carries sediment eroded from extensive reddish soils like Port Silt Loam in Oklahoma. The Yellow River in China carries yellow sediment from eroding loess soils. Mollisols in the Great Plains of North America are darkened and enriched by organic matter. Podsols in boreal forests have highly contrasting layers due to acidity and leaching.

In general, color is determined by the organic matter content, drainage conditions, and degree of oxidation. Soil color, while easily discerned, has little use in predicting soil characteristics. It is of use in distinguishing boundaries within a soil profile, determining the origin of a soil›s parent material, as an indication of wetness and waterlogged conditions, and as a qualitative means of measuring organic, salt and carbonate contents of soils. Color is recorded in the Munsell color system as for instance 10YR3/4 Dusky Red.

Soil color is primarily influenced by soil mineralogy. Many soil colours are due to various iron minerals. The development and distribution of colour in a soil profile result from chemical and biological weathering, especially redox reactions. As the primary minerals in soil parent material weather, the elements combine into new and colourful compounds. Iron forms secondary minerals of a yellow or red colour, organic matter decomposes into black and brown compounds, and manganese, sulfur and nitrogen can form black mineral deposits. These pigments can produce various colour patterns within a soil. Aerobic conditions produce uniform or gradual colour changes, while reducing environments (anaerobic) result in rapid colour flow with complex, mottled patterns and points of colour concentration.

Resistivity

Soil resistivity is a measure of a soil's ability to retard the conduction of an electric current. The electrical resistivity of soil can affect the rate of galvanic corrosion of metallic structures in contact with the soil. Higher moisture content or increased electrolyte concentration can lower resistivity and increase conductivity, thereby increasing the rate of corrosion. Soil resistivity values typically range from about 2 to 1000 Ω·m, but more extreme values are not unusual.

Soil Water

Water that enters a field is removed from a field by runoff, drainage, evaporation or transpiration. Runoff is the water that flows on the surface to the edge of the field; drainage is the water that flows through the soil downward or toward the edge of the field underground; evaporative water loss from a field is that part of the water that evaporates into the atmosphere directly from the field's surface; transpiration is the loss of water from the field by its evaporation from the plant itself.

Water affects soil formation, structure, stability and erosion but is of primary concern with respect to plant growth. Water is essential to plants for four reasons:

- It constitutes 80%-95% of the plant's protoplasm.
- It is essential for photosynthesis.
- It is the solvent in which nutrients are carried to, into and throughout the plant.
- It provides the turgidity by which the plant keeps itself in proper position.

In addition, water alters the soil profile by dissolving and re-depositing minerals, often at lower levels, and possibly leaving the soil sterile in the case of extreme rainfall and drainage. In a loam soil, solids constitute half the volume, gas one-quarter of the volume, and water one-quarter of the volume of which only half of which will be available to most plants.

A flooded field will drain the gravitational water under the influence of gravity until water's adhesive and cohesive forces resist further drainage at which point it is said to have reached field capacity. At that point, plants must apply suction to draw water from a soil. The water that plants may draw from the soil is called the available water. Once the available water is used up the remaining moisture is called unavailable water as the plant cannot produce sufficient suction to draw that water in. A plant must produce suction that increases from zero for a flooded field to 1/3 bar at field dry condition (one bar is a little less than one atmosphere pressure). At 15 bar suction, wilting percent, seeds will not germinate, plants begin to wilt and then die. Water moves in soil under the influence of gravity, osmosis and capillarity. When water enters the soil, it displaces air from some of the pores, since air content of a soil is inversely related to its water content.

The rate at which a soil can absorb water depends on the soil and its other conditions. As a plant grows, its roots remove water from the largest pores first. Soon the larger pores hold only air, and the remaining water is found only in the intermediate- and smallest-sized pores. The water in the smallest pores is so strongly held to particle surfaces that plant roots cannot pull it away. Consequently, not all soil water is available to plants. When saturated, the soil may lose nutrients as the water drains. Water moves in a draining field under the influence of pressure where the soil is locally saturated and by capillarity pull to dryer parts of the soil. Most plant water needs are supplied from the suction caused by evaporation from plant leaves and 10% is supplied by "suction" created by osmotic pressure differences between the plant interior and the soil water. Plant roots must seek out water. Insufficient water will damage the yield of a crop. Most of the available water is used in transpiration to pull nutrients into the plant.

Water Retention Forces

Water is retained in a soil when the adhesive force of attraction that water's hydrogen atoms have for the oxygen of soil particles is stronger than the cohesive forces that water's hydrogen feels for other water oxygen atoms. When a field is flooded, the soil pore space is completely filled by water. The field will drain under the force of gravity until it reaches what is called field capacity, at which point the smallest pores are filled with water and the largest with water and gases.The total amount of water held when field capacity is reached is a function of the specific surface area of the soil particles. As a result, high clay and high organic soils have higher field capacities. The total force required to pull or push water out of soil is termed suction and usually expressed in units of bars (10^5 pascal) which is just a little less than one-atmosphere pressure. Alternatively, the terms "tension" or "moisture potential" may be used.

Moisture Classification

The forces with which water is held in soils determine its availability to plants. Forces of adhesion hold water strongly to mineral and humus surfaces and less strongly to itself by cohesive forces. A plant's root may penetrate a very small volume of water that is adhering to soil and be initially able to draw in water that is only lightly held by the cohesive forces. But as the droplet is drawn down, the forces of adhesion of the water for the soil particles make reducing the volume of water increasingly difficult until the plant cannot produce sufficient suction to use the remaining water. The remaining water is considered unavailable. The amount of available water depends upon the soil texture and humus amounts and the type of plant attempting to use the water. Cacti, for example, can produce greater suction than can agricultural crop plants.

The following description applies to a loam soil and agricultural crops. When a field is flooded, it is said to be saturated and all available air space is occupied by water. The suction required to draw water into a plant root is zero. As the field drains under the influence of gravity (drained water is called gravitational water or drain-able water), the suction a plant must produce to use such water increases to 1/3 bar. At that point, the soil is said to have reached field capacity, and plants that use the water must produce increasingly higher suction, finally up to 15 bar. At 15 bar suction, the soil water amount is called wilting percent. At that suction the plant cannot sustain its water needs as water is still being lost from the plant by transpiration; the plant's turgidity is lost, and it wilts. The next level, called air-dry, occurs at 1000 bar suction. Finally the oven dry condition is reached at 10,000 bar suction. All water below wilting percentage is called unavailable water.

Soil Moisture Content

When the soil moisture content is optimal for plant growth, the water in the large and intermediate size pores can move about in the soil and be easily used by plants. The amount of water remaining in a soil drained to field capacity and the amount that is available are functions of the soil type. Sandy soil will retain very little water, while clay will hold the maximum amount. The time required to drain a field from flooded condition for a clay loam that begins at 43% water by weight to a field capacity of 21.5% is six days, whereas a sand loam that is flooded to its maximum of 22% water will take two days to reach field capacity of 11.3% water. The available water for the clay loam might be 11.3% whereas for the sand loam it might be only 7.9% by weight.

Wilting point, field capacity, and available water capacity of various soil textures

Soil Texture	Wilting Point		Field Capacity		Available water capacity	
	Water per foot of soil depth		**Water per foot of soil depth**		**Water per foot of soil depth**	
	%	**in.**	**%**	**in.**	**%**	**in.**
Medium sand	1.7	0.3	6.8	1.2	5.1	0.9
Fine sand	2.3	0.4	8.5	1.5	6.2	1.1
Sandy loam	3.4	0.6	11.3	2.0	7.9	1.4
Fine sandy loam	4.5	0.8	14.7	2.6	10.2	1.8
Loam	6.8	1.2	18.1	3.2	11.3	2.0
Silt loam	7.9	1.4	19.8	3.5	11.9	2.1
Clay loam	10.2	1.8	21.5	3.8	11.3	2.0
Clay	14.7	2.6	22.6	4.0	7.9	1.4

The above are average values for the soil textures as the percentages of sand, silt and clay vary.

Water Flow in Soils

Water moves through soil due to the force of gravity, osmosis and capillarity. At zero to one-third bar suction, water is pushed through soil from the point of its application under the force of gravity and the pressure gradient created by the pressure of the water; this is called saturated flow. At higher suction, water movement is pulled by capillarity from wetter toward dryer soil. This is caused by water's adhesion to soil solids, and is called unsaturated flow.

Water infiltration and movement in soil is controlled by six factors:

- Soil texture
- Soil structure. Fine-textured soils with granular structure are most favourable to infiltration of water.
- The amount of organic matter. Coarse matter is best and if on the surface helps prevent the destruction of soil structure and the creation of crusts.
- Depth of soil to impervious layers such as hardpans or bedrock
- The amount of water already in the soil
- Soil temperature. Warm soils take in water faster while frozen soils may not be able to absorb depending on the type of freezing.

Water infiltration rates range from 0.25 cm (0.098 in) per hour for high clay soils to 2.5 cm (0.98 in) per hour for sand and well stabilised and aggregated soil structures. Water flows through the ground unevenly, called "gravity fingers", because of the surface tension between water particles. Tree roots create paths for rainwater flow through soil by breaking though soil including clay layers: one study showed roots increasing infiltration of water by 153% and another study showed an increase by 27 times. Flooding temporarily increases soil permeability in river beds, helping to recharge aquifers.

Saturated Flow

Water applied to a soil is pushed by pressure gradients from the point of its application where it is saturated locally, to less saturated areas. Once soil is completely wetted, any more water will move downward, or percolate, carrying with it clay, humus and nutrients, primarily cations, out of the range of plant roots. In order of decreasing solubility, the leached nutrients are:

- Calcium
- Magnesium, Sulfur, Potassium; depending upon soil composition
- Nitrogen; usually little, unless nitrate fertiliser was applied recently
- Phosphorus; very little as its forms in soil are of low solubility.

In the United States percolation water due to rainfall ranges from zero inches just east of the Rocky Mountains to twenty or more inches in the Appalachian Mountains and the north coast of the Gulf of Mexico.

Unsaturated Flow

At suctions less than one-third bar, water moves in all directions via unsaturated flow at a rate that is dependent on the square of the diameter of the water-filled pores. Water is pulled by capillary action due to the adhesion force of water to the soil solids, producing a suction gradient from wet towards drier soil. Doubling the diameter of the pores increases the flow rate by a factor of four. Large pores drained by gravity and not filled with water do not greatly increase the flow rate for unsaturated flow. Water flow is primarily from coarse-textured soil into fine-textured soil and is slowest in fine-textured soils such as clay.

Water Uptake by Plants

Of equal importance to the storage and movement of water in soil is the means by which plants acquire it and their nutrients. Ninety percent of water is taken up by plants as passive absorption caused by the pulling force of water evaporating (transpiring) from the long column of water that leads from the plant's roots to its leaves. In addition, the high concentration of salts within plant roots creates an osmotic pressure gradient that pushes soil water into the roots. Osmotic absorption becomes more important during times of low water transpiration caused by lower temperatures (for example at night) or high humidity. It is the process that causes guttation.

Root extension is vital for plant survival. A study of a single winter rye plant grown for four months in one cubic foot of loam soil showed that the plant developed 13,800,000 roots, a total of 385 miles in length with 2,550 square feet in surface area; and 14 billion hair roots of 6,600 miles total length and 4,320 square feet total area; for a total surface area of 6,870 square feet (83 ft squared). The total surface area of the loam soil was estimated to be 560,000 square feet. In other words, the roots were in contact with only 1.2% of the soil.

Roots must seek out water as the unsaturated flow of water in soil can move only at a rate of up to 2.5 cm (onc inch) per day; as a result they are constantly dying and growing as they seek out high concentrations of soil moisture. Insufficient soil moisture, to the point of causing wilting, will cause permanent damage and crop yields will suffer. When grain sorghum was exposed to soil suction as low as 13.0 bar during the seed head emergence through bloom and seed set stages of growth, its production was reduced by 34%.

CONSUMPTIVE USE AND WATER EFFICIENCY

Only a small fraction (0.1% to 1%) of the water used by a plant is held within the plant. The majority is ultimately lost via transpiration, while evaporation from the soil surface is also substantial. Transpiration plus evaporative soil moisture loss is called evapotranspiration. Evapotranspiration plus water held in the plant totals to consumptive use, which is nearly identical to evapotranspiration.

The total water used in an agricultural field includes runoff, drainage and consumptive use. The use of loose mulches will reduce evaporative losses for a period after a field is irrigated, but in the end the total evaporative loss will approach that of an uncovered soil. The benefit from mulch is to keep the moisture available during the seedling stage. Water use efficiency is measured by transpiration ratio, which is the ratio of the total water transpired by a plant to the dry weight of the harvested plant. Transpiration ratios for crops range from 300 to 700. For example, alfalfa may have a transpiration ratio of 500 and as a result 500 kilograms of water will produce one kilogram of dry alfalfa.

Composition of Soil Particles

Soil particles can be classified by their chemical composition (mineralogy) as well as their size. The particle size distribution of a soil, its texture, determines many of the properties of that soil, but the mineralogy of those particles can strongly modify those properties. The mineralogy of the finest soil particles, clay, is especially important.

Gravel, Sand and Silt

Gravel, sand and silt are the larger soil particles, and their mineralogy is often inherited from the parent material of the soil, but may include products of weathering (such as concretions of calcium carbonate or iron oxide), or residues of plant and animal life (such as silica phytoliths). Quartz is the most common mineral in the sand or silt fraction as it is resistant to chemical weathering; other common minerals are feldspars, micas and ferromagnesian minerals such as pyroxenes, amphiboles and olivines.

Mineral Colloids; Soil Clays

Due to its high specific surface area and its unbalanced negative charges, clay is the most active mineral component of soil. It is a colloidal and most often a crystalline material. In soils, clay is a soil textural class and is defined in a physical sense as any mineral particle less than 2 μm (8×10^{-5} in) in effective diameter. Many soil minerals, such as gypsum, carbonates, or quartz, are small enough to be classified as clay based on their physical size, but chemically they do not afford the same utility as do clay minerals. Chemically, clay is a range of minerals with certain reactive properties.

Clay was once thought to be very small particles of quartz, feldspar, mica, hornblende or augite, but it is now known to be (with the exception of mica-based clays) a precipitate with a mineralogical composition that is dependent on but different from its parent materials and is classed as a secondary mineral. The type of clay that is formed is a function of the parent material and the composition of the minerals in solution. Clay minerals continue to be formed as long as the soil exists.Mica-based clays result from a modification of the primary mica mineral in such a way that it behaves and is classed as a clay. Most clays are crystalline, but some are amorphous. The clays of a soil are a mixture of the various types of clay, but one type predominates.

There are four groups of clay: layer silicates; crystalline chain silicates; metal oxides and hydroxides and oxy-oxides; and amorphous; and allophanes. Most clays are crystalline and most are made up of three or four planes of oxygen held together by planes of aluminium and silicon by way of ionic bonds that together form a single layer of clay. The spatial arrangement of the oxygen atoms determines clay›s structure. Half of the weight of clay is oxygen, but on a volume basis oxygen is ninety percent. The layers of clay are sometimes held together through hydrogen bonds or potassium bridges and as a result will swell less in the presence of water. Other clays, such as montmorillonite, have layers that are loosely attached and will swell greatly when water intervenes between the layers.

There are four groups of clays:

- Layer Crystalline alumino-silica clays: montmorillonite, illite, vermiculite, chlorite, kaolinite.
- Crystalline Chain carbonate and sulfate minerals: calcite ($CaCO_3$), dolomite ($CaMg(CO_3)_2$) and gypsum ($CaSO_4 \cdot 2H2O$).
- Amorphous clays: young mixtures of silica (SiO_2-OH) and alumina ($Al(OH)_3$) which have not had time to form regular crystals.
- Sesquioxide clays: old, highly leached clays which result in oxides of iron, aluminium and titanium.

Alumino-Silica Clays

Alumino-silica clays are characterised by their regular crystalline structure. Oxygen in ionic bonds with silicon forms a tetrahedral coordination (silicon at the center) which in turn forms sheets of silica. Two sheets of silica are bonded together by a plane of aluminium which forms an octahedral coordination, called alumina, with the oxygens of the silica sheet above and that below it. Hydroxyl ions (OH^-) sometimes substitute for oxygen. During the clay formation process, Al^{3+}may substitute for Si^{4+} in the silica layer, and as much as one fourth of the aluminium Al^{3+} may be substituted by Zn^{2+}, Mg^{2+} or Fe^{2+} in the alumina layer. The substitution of lower-valence cations for higher-valence cations (isomorphous substitution) gives clay a local negative charge on an oxygen atom that attracts and holds water and positively charged soil cations, some of which are of value for plant growth. Isomorphous substitution occurs during the clay›s formation and does not change with time.

- **Montmorillonite** clay is made of four planes of oxygen with two silicon and one central aluminium plane intervening. The alumino-silicate montmorillonite clay is said to have a 2:1 ratio of silicon to aluminium. The seven planes together form a single crystal of montmorillonite. The crystals are weakly held together and water may intervene, causing the clay to swell up to ten times its dry volume. It occurs in soils which have had little leaching, hence it is found in arid regions. As the crystals are not bonded face to face, the entire surface is exposed and available for surface reactions, hence it has a high cation exchange capacity (CEC).
- **Illite** is a 2:1 clay similar in structure to montmorillonite but has potassium bridges between the faces of the clay crystals and the degree of swelling depends on the degree of weathering of the potassium. The active surface area is reduced due to the potassium bonds. Illite originates from the modification of mica, a primary mineral. It is often found together with montmorillonite and

its primary minerals. It has moderate CEC.

- **Vermiculite** is a mica-based clay similar to illite, but the crystals of clay are held together more loosely by hydrated magnesium and it will swell, but not as much as does montmorillonite. It has very high CEC.
- **Chlorite** is similar to vermiculite, but the loose bonding by occasional hydrated magnesium, as in vermiculite, is replaced by a hydrated magnesium sheet, that firmly bonds the planes above and below it. It has two planes of silicon, one of aluminium and one of magnesium; hence it is a 2:2 clay. Chlorite does not swell and it has low CEC.
- **Kaolinite** is very common, highly weathered clay, and more common than montmorillonite in acid soils. It has one silica and one alumina plane per crystal; hence it is a 1:1 type clay. One plane of silica of montmorillonite is dissolved and is replaced with hydroxyls, which produces strong hydrogen bonds to the oxygen in the next crystal of clay. As a result, kaolinite does not swell in water and has a low specific surface area, and as almost no isomorphous substitution has occurred it has a low CEC. Where rainfall is high, acid soils selectively leach more silica than alumina from the original clays, leaving kaolinite. Even heavier weathering results in sesquioxide clays.

silica-sesquioxide

Crystalline Chain Clays

The carbonate and sulfate minerals are much more soluble and hence are found primarily in desert soils where leaching is less active.

Amorphous Clays

Amorphous clays are young, and commonly found in volcanic ash. They are mixtures of alumina and silica which have not formed the ordered crystal shape of alumino-silica clays which time would provide. The majority of their negative charges originates from hydroxyl ions, which can gain or lose a hydrogen ion (H^+) in response to soil pH, in such way was as to buffer the soil pH. They may have either a negative charge provided by the attached hydroxyl ion (OH^-), which can attract a cation, or lose the hydrogen of the hydroxyl to solution and display a positive charge which can attract anions. As a result, they may display either high CEC in an acid soil solution, or high anion exchange capacity in a basic soil solution.

Sesquioxide Clays

Sesquioxide **clays** are a product of heavy rainfall that has leached most of the silica from alumino-silica clay, leaving the less soluble oxides iron hematite (Fe_2O_3), iron hydroxide ($Fe(OH)_3$), aluminium hydroxide gibbsite ($Al(OH)_3$), hydrated manganese birnessite (MnO_2). It takes hundreds of thousands of years of leaching to create sesquioxide clays. Sesqui is Latin for "one and one-half": there are three parts oxygen to two parts iron or aluminium; hence the ratio is one and one-half (not true for all). They are hydrated and act as either amorphous or crystalline. They are not sticky and do not swell, and soils high in them behave much like sand and can rapidly pass water. They are able to hold large quantities of phosphates. Sesquioxides have low CEC but are able to hold anions as well as cations. Such soils range from yellow to red in colour. Such clays tend to hold phosphorus so tightly that it is unavailable for absorption by plants.

Organic Colloids

Humus is the final state of decomposition of organic matter. While it may linger for a thousand years, on the larger scale of the age of the mineral soil components, it is temporary. It is composed of the very stable lignins (30%) and complex sugars (polyuronides, 30%), proteins (30%), waxes, and fats that are resistant to breakdown by microbes. Its chemical assay is 60% carbon, 5% nitrogen, some oxygen and the remainder hydrogen, sulfur, and phosphorus. On a dry weight basis, the CEC of humus is many times greater than that of clay.

Carbon and Terra Preta

In the extreme environment of high temperatures and the leaching caused by the heavy rain of tropical rain forests, the clay and organic colloids are largely destroyed. The heavy rains wash the alumino-silicate clays from the soil leaving only sesquioxide clays of low CEC. The high temperatures and humidity allow bacteria and fungi to virtually dissolve any organic matter on the rain-forest floor overnight and much of the nutrients are volatilized or leached from the soil and lost. However, carbon in the form of charcoal is far more stable than soil colloids and is capable of performing many of the functions of the soil colloids of sub-tropical soils. Soil containing substantial quantities of charcoal, of an anthropogenic origin, is called terra preta. Research into terra preta is still young but is promising. Fallow periods "on the Amazonian Dark Earths can be as short as 6 months, whereas fallow periods on oxisols are usually 8 to 10 years long".

MICROMORPHOLOGY

While soil micromorphology begins in the field with the routine and careful use of a 10x hand lens, much more can be described by careful description of thin sections made of the soil with the aid of a petrographic polarizing light microscope. The soil can be impregnated with an epoxy resin, but more commonly with a polyester resin (crystic 17449) and sliced and ground to 0.03 millimeter thickness and examined by passing light through the thin soil plasma.

Porosity

Porosity of topsoil typically decreases as grain size increases. This is due to soil aggregate formation in finer textured surface soils when subject to soil biological processes. Aggregation involves particulate adhesion and higher resistance to compaction. Typical bulk density of sandy soil is between 1.5 and 1.7 g/cm^3. This calculates to a porosity between 0.43 and 0.36. Typical bulk density of clay soil is between 1.1 and 1.3 g/cm^3. This calculates to a porosity between 0.58 and 0.51. This seems counterintuitive because clay soils are termed heavy, implying lower porosity. Heavy apparently refers to a gravitational moisture content effect in combination with terminology that harkens back to the relative force required to pull a tillage implement through the clayey soil at field moisture content as compared to sand.

Porosity of subsurface soil is lower than in surface soil due to compaction by gravity. Porosity of 0.20 is considered normal for unsorted gravel size material at depths below the biomantle. Porosity in finer material below the aggregating influence of pedogenesis can be expected to approximate this value.

Soil porosity is complex. Traditional models regard porosity as continuous. This fails to account for anomalous features and produces only approximate results. Furthermore it cannot help model the influence of environmental factors which affect pore geometry. A number of more complex models have been proposed, including fractals, bubble theory, cracking theory, Boolean grain process, packed sphere, and numerous other models.

SOIL MECHANICS

Soil mechanics is a branch of soil physics and engineering mechanics that describes the behavior of soils. It differs from fluid mechanics and solid mechanics in the sense that soils consist of a heterogeneous mixture of fluids (usually air and water) and particles (usually clay, silt, sand, and gravel) but soil may also contain organic solids and other matter. Along with rock mechanics, soil mechanics provides the theoretical basis for analysis in geotechnical engineering, a subdiscipline of civil engineering, and engineering geology, a subdiscipline of geology. Soil mechanics is used to analyze the deformations of and flow of fluids within natural and man-made structures that are supported on or made of soil, or structures that are buried in soils. Example applications are building and bridge foundations, retaining walls, dams, and buried pipeline systems. Principles of soil mechanics are also used in related disciplines such as engineering geology, geophysical engineering, coastal engineering, agricultural engineering, hydrology and soil physics.

The Tower of Pisa – an example of a problem due to deformation of soil.

This article describes the genesis and composition of soil, the distinction between pore water pressure and inter-granular effective stress, capillary action of fluids in the soil pore spaces, soil classification, seepage and permeability, time dependent change of volume due to squeezing water out of tiny pore spaces, also known as consolidation, shear strength and stiffness of soils. The shear strength of soils is primarily derived from friction between the particles and interlocking, which are very sensitive to the effective stress. The article concludes with some examples of applications of the principles of soil mechanics such as slope stability, lateral earth pressure on retaining walls, and bearing capacity of foundations.

Slope instability issues for a temporary flood control levee in North Dakota, 2009

Earthwork in Germany

Fox Glacier, New Zealand: Soil produced and transported by intense weathering and erosion.

GENESIS AND COMPOSITION OF SOILS

Genesis

The primary mechanism of soil creation is the weathering of rock. All rock types (igneous rock, metamorphic rock and sedimentary rock) may be broken down into small particles to create soil. Weathering mechanisms are physical weathering, chemical weathering, and biological weathering Human activities such as excavation, blasting, and waste disposal, may also create soil. Over geologic time, deeply buried soils may be altered by pressure and temperature to become metamorphic or sedimentary rock, and if melted and solidified again, they would complete the geologic cycle by becoming igneous rock.

Physical weathering includes temperature effects, freeze and thaw of water in cracks, rain, wind, impact and other mechanisms. Chemical weathering includes dissolution of matter composing a rock and precipitation in the form of another mineral. Clay minerals, for example can be formed by weathering of feldspar, which is the most common mineral present in igneous rock.

The most common mineral constituent of silt and sand is quartz, also called silica, which has the chemical name silicon dioxide. The reason that feldspar is most common in rocks but silicon is more prevalent in soils is that feldspar is much more soluble than silica.

Silt, Sand, and Gravel are basically little pieces of broken rocks.

According to the Unified Soil Classification System, silt particle sizes are in the range of 0.002 mm to 0.075 mm and sand particles have sizes in the range of 0.075 mm to 4.75 mm.

Gravel particles are broken pieces of rock in the size range 4.75 mm to 100 mm.

Particles larger than gravel are called cobbles and boulders.

Transport

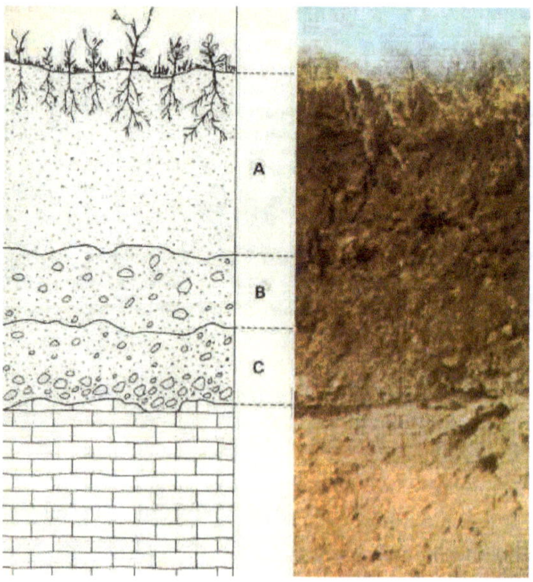

Example soil horizons. a) top soil and colluvium b) mature residual soil c) young residual soil d) weathered rock.

Soil deposits are affected by the mechanism of transport and deposition to their location. Soils that are not transported are called residual soils—they exist at the same location as the rock from which they were generated. Decomposed granite is a common example of a residual soil. The common mechanisms of transport are the actions of gravity, ice, water, and wind. Wind blown soils include dune sands and loess. Water carries particles of different size depending on the speed of the water, thus soils transported by water are graded according to their size. Silt and clay may settle out in a lake, and gravel and sand collect at the bottom of a river bed. Wind blown soil deposits (aeolian soils) also tend to be sorted according to their grain size. Erosion at the base of glaciers is powerful enough to pick up large rocks and boulders as well as soil; soils dropped by melting ice can be a well graded mixture of widely varying particle sizes. Gravity on its own may also carry particles down from the top of a mountain to make a pile of soil and boulders at the base; soil deposits transported by gravity are called colluvium.

The mechanism of transport also has a major effect on the particle shape. For example, low velocity grinding in a river bed will produce rounded particles. Freshly fractured colluvium particles often have a very angular shape.

Soil Composition

Soil mineralogy

Silts, sands and gravels are classified by their size, and hence they may consist of a variety of minerals. Owing to the stability of quartz compared to other rock minerals, quartz is the most common constituent of sand and silt . Mica, and feldspar are other common minerals present in sands and silts. The mineral constituents of gravel may be more similar to that of the parent rock.

The common clay minerals are montmorillonite or smectite, illite, and kaolinite or kaolin. These minerals tend to form in sheet or plate like structures, with length typically ranging between 10^{-7} m and $4x10^{-6}$ m and thickness typically ranging between 10^{-9} m and $2x10^{-6}$ m, and they have a relatively large specific surface area. The specific surface area (SSA) is defined as the ratio of the surface area of particles to the mass of the particles. Clay minerals typically have specific surface areas in the range of 10 to 1,000 square meters per gram of solid. Due to the large surface area available for chemical, electrostatic, and van der Waals interaction, the mechanical behavior of clay minerals is very sensitive to the amount of pore fluid available and the type and amount of dissolved ions in the pore fluid. To anticipate the effect of clay on the way a soil will behave, it is necessary to know the kinds of clays as well as the amount present. As home builders and highway engineers know all too well, soils containing certain high-activity clays make very unstable material on which to build because they swell when wet and shrink when dry. This shrink-and-swell action can easily crack foundations and cause retaining walls to collapse. These clays also become extremely sticky and difficult to work with when they are wet. In contrast, low-activity clays, formed under different conditions, can be very stable and easy to work with. The minerals of soils are predominantly formed by atoms of oxygen, silicon, hydrogen, and aluminum, organized in various crystalline forms. These elements along with calcium, sodium, potassium, magnesium, and carbon constitute over 99 per cent of the solid mass of soils.

Grain size Distribution

Soils consist of a mixture of particles of different size, shape and mineralogy. Because the size of the particles obviously has a significant effect on the soil behavior, the grain size and grain size distribution are used to classify soils. The grain size distribution describes the relative proportions of particles of various sizes. The grain size is often visualized in a cumulative distribution graph which, for example, plots the percentage of particles finer than a given size as a function of size. The median grain size, {\displaystyle D_{50}}, is the size for which 50% of the particle mass consists of finer particles. Soil behavior, especially the hydraulic conductivity, tends to be dominated by the smaller particles, hence, the term "effective size", denoted by {\displaystyle D_{10}}, is defined as the size for which 10% of the particle mass consists of finer particles.

Sands and gravels that possess a wide range of particle sizes with a smooth distribution of particle sizes are called well graded soils. If the soil particles in a sample are predominantly in a relatively narrow range of sizes, the sample is uniformly graded. If a soil sample has distinct gaps in the gradation curve, e.g., a mixture of gravel and fine sand, with no coarse sand, the sample may be gap graded. Uniformly graded and gap graded soils are both considered to be poorly graded. There are many methods for measuring particle size distribution. The two traditional methods are sieve analysis and hydrometer analysis.

Sieve Analysis

Sieve

The size distribution of gravel and sand particles are typically measured using sieve analysis. The formal procedure is described in ASTM D6913-04(2009). A stack of sieves with accurately dimensioned holes between a mesh of wires is used to separate the particles into size bins. A known volume of dried soil, with clods broken down to individual particles, is put into the top of a stack of sieves arranged from coarse to fine. The stack of sieves is shaken for a standard period of time so that the particles are sorted into size bins. This method works reasonably well for particles in the sand and gravel size range. Fine particles tend to stick to each other, and hence the sieving process is not an effective method. If there are a lot of fines (silt and clay) present in the soil it may be necessary to run water through the sieves to wash the coarse particles and clods through.

A variety of sieve sizes are available. The boundary between sand and silt is arbitrary. According to the Unified Soil Classification System, a #4 sieve (4 openings per inch) having 4.75mm opening size separates sand from gravel and a #200 sieve with an 0.075 mm opening separates sand from silt and clay. According to the British standard, 0.063 mm is the boundary between sand and silt, and 2 mm is the boundary between sand and gravel.

Hydrometer Analysis

The classification of fine-grained soils, i.e., soils that are finer than sand, is determined primarily by their Atterberg limits, not by their grain size. If it is important to determine the grain size distribution of fine-grained soils, the hydrometer test may be performed. In the hydrometer tests, the soil particles are mixed with water and shaken to produce a dilute suspension in a glass cylinder, and then the cylinder is left to sit. A hydrometer is used to measure the density of the suspension as a function of time. Clay particles may take several hours to settle past the depth of measurement of the hydrometer. Sand particles may take less than a second. Stoke's law provides the theoretical basis to calculate the relationship between sedimentation velocity and particle size. ASTM provides the detailed procedures for performing the Hydrometer test.

Clay particles can be sufficiently small that they never settle because they are kept in suspension by Brownian motion, in which case they may be classified as colloids.

Mass-volume Relations

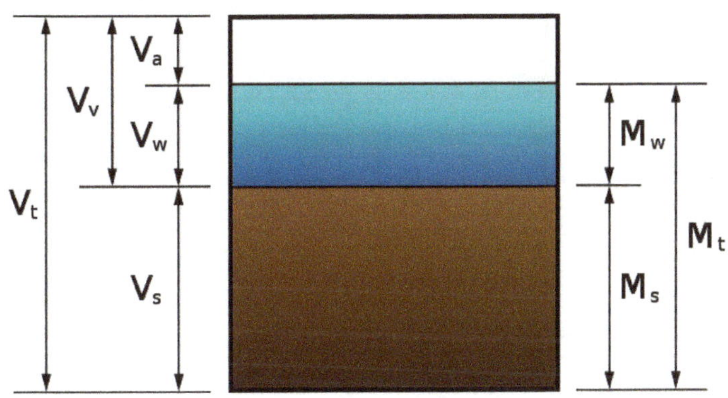

A phase diagram of soil indicating the masses and volumes of air, solid, water, and voids.

There are a variety of parameters used to describe the relative proportions of air, water and solid in a soil. This section defines these parameters and some of their interrelationships. The basic notation is as follows:

M_a, M_w, and Ms represent the volumes of air, water and solids in a soil mixture;

W_a, W_w, and W_s represent the weights of air, water and solids in a soil mixture;

M_a, M_w, and Ms represent the masses of air, water and solids in a soil mixture;

ρ_a, ρ_w, and ρ_s represent the densities of the constituents (air, water and solids) in a soil mixture;

Note that the weights, W, can be obtained by multiplying the mass, M, by the acceleration due to gravity, g; e.g., $W_s = M_s g$

Specific Gravity is the ratio of the density of one material compared to the density of pure water ($\rho_w = 1g/cm^3$).

Specific gravity of solids, $G_s = \frac{\rho_s}{\rho_w}$

Note that unit weights, conventionally denoted by the symbol γ may be obtained by multiplying the density (ρ) of a material by the acceleration due to gravity, g .

Density, **Bulk Density**, or **Wet Density**, ρ, are different names for the density of the mixture, i.e., the total mass of air, water, solids divided by the total volume of air water and solids (the mass of air is assumed to be zero for practical purposes):

$$\rho = \frac{M_s + M_w}{V_s + V_w + V_a} = \frac{M_t}{V_t}$$

Dry Density, ρ_d, is the mass of solids divided by the total volume of air water and solids:

$$\rho_d = \frac{M_s}{V_s + V_w + V_a} = \frac{M_s}{V_t}$$

Buoyant Density, ρ' , defined as the density of the mixture minus the density of water is useful if the soil is submerged under water:

$$\rho' = \rho - \rho_w$$

where ρ_w is the density of water

Water Content, w is the ratio of mass of water to mass of solid. It is easily measured by weighing a sample of the soil, drying it out in an oven and re-weighing. Standard procedures are described by ASTM.

$$w = \frac{M_w}{M_s} = \frac{W_w}{W_s}$$

Void ratio, e, is the ratio of the volume of voids to the volume of solids:

$$e = \frac{V_V}{V_S} = \frac{V_V}{V_T - V_V} = \frac{n}{1-n}$$

Porosity, n, is the ratio of volume of voids to the total volume, and is related to the void ratio:

$$n = \frac{V_v}{V_t} = \frac{V_v}{V_s + V_v} = \frac{e}{1+e}$$

Degree of saturation, S , is the ratio of the volume of water to the volume of voids:

$$S = \frac{V_w}{V_v}$$

From the above definitions, some useful relationships can be derived by use of basic algebra.

$$\rho = \frac{(G_s + Se)\rho_w}{1+e}$$

$$\rho = \frac{(1+w)G_s\rho_w}{1+e}$$

$$w = \frac{Se}{G_s}$$

EFFECTIVE STRESS AND CAPILLARITY: HYDROSTATIC CONDITIONS

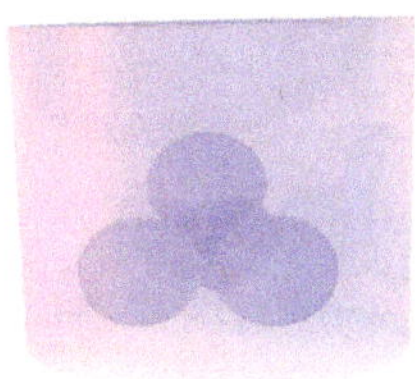

Spheres immersed in water, reducing effective stress.

To understand the mechanics of soils it is necessary to understand how normal stresses and shear stresses are shared by the different phases. Neither gas nor liquid provide significant resistance to shear stress. The shear resistance of soil is provided by friction and interlocking of the particles. The friction depends on the intergranular contact stresses between solid particles. The normal stresses, on the other hand, are shared by the fluid and the particles. Although the pore air is relatively compressible, and hence takes little normal stress in most geotechnical problems, liquid water is relatively incompressible and if the voids are saturated with water, the pore water must be squeezed out in order to pack the particles closer together.

The principle of effective stress, introduced by Karl Terzaghi, states that the effective stress σ' (i.e., the average intergranular stress between solid particles) may be calculated by a simple subtraction of the pore pressure from the total stress:

$$\sigma' = \sigma - u$$

where σ is the total stress and u is the pore pressure. It is not practical to measure σ› directly, so in practice the vertical effective stress is calculated from the pore pressure and vertical total stress. The distinction between the terms pressure and stress is also important. By definition, pressure at a point is equal in all directions but stresses at a point can be different in different directions. In soil mechanics, compressive stresses and pressures are considered to be positive and tensile stresses are considered to be negative, which is different from the solid mechanics sign convention for stress.

Total Stress

For level ground conditions, the total vertical stress at a point, σ_v, on average, is the weight of everything above that point per unit area. The vertical stress beneath a uniform surface layer with density ρ, and thickness H is for example:

$$\sigma_v = \rho g H = \gamma H$$

where g is the acceleration due to gravity, and γ is the unit weight of the overlying layer. If there are multiple layers of soil or water above the point of interest, the vertical stress may be calculated by summing the product of the unit weight and thickness of all of the overlying layers. Total stress increases with increasing depth in proportion to the density of the overlying soil.

It is not possible to calculate the horizontal total stress in this way. Lateral earth pressures are addressed elsewhere.

Pore Water Pressure

Hydrostatic conditions

Water is drawn into a small tube by surface tension. Water pressure, u, is negative above and positive below the free water surface

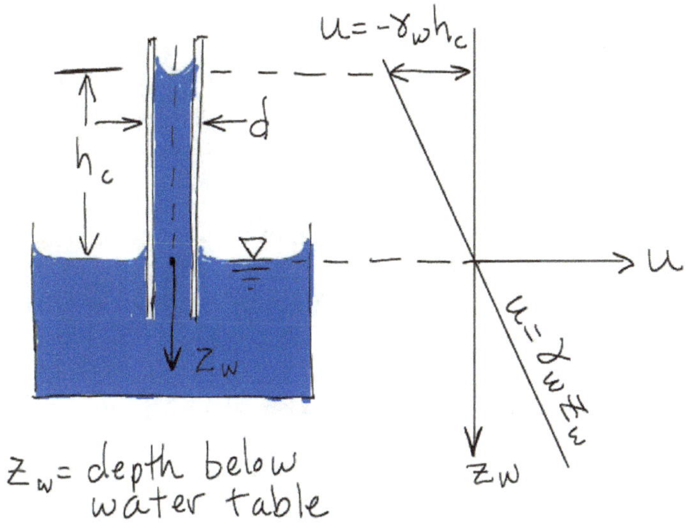

If there is no pore water flow occurring in the soil, the pore water pressures will be hydrostatic. The water table is located at the depth where the water pressure is equal to the atmospheric pressure. For hydrostatic conditions, the water pressure increases linearly with depth below the water table:

$$u = \rho_w g z_w$$

where ρ_w is the density of water, and z_w is the depth below the water table.

Capillary Action

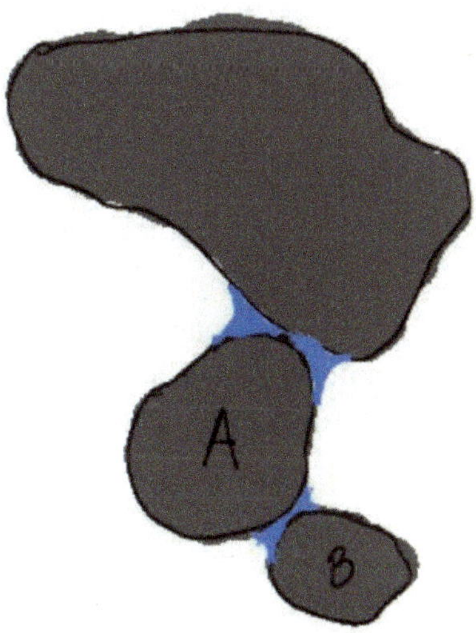

Water at particle contacts

Due to surface tension, water will rise up in a small capillary tube above a free surface of water. Likewise, water will rise up above the water table into the small pore spaces around the soil particles. In fact the soil may be completely saturated for some distance above the water table. Above the height of capillary saturation, the soil may be wet but the water content will decrease with elevation. If the water in the capillary zone is not moving, the water pressure obeys the equation of hydrostatic equilibrium, $u = \rho_w g z_w$, but note that z_w, is negative above the water table. Hence, hydrostatic water pressures are negative above the water table. The thickness of the zone of capillary saturation depends on the pore size, but typically, the heights vary between a centimeter or so for coarse sand to tens of meters for a silt or clay. In fact the pore space of soil is a uniform fractal e.g. a set of uniformly distributed D-dimensional fractals of average linear size L. For the clay soil it has been found that L=0.15 mm and D=2.7.

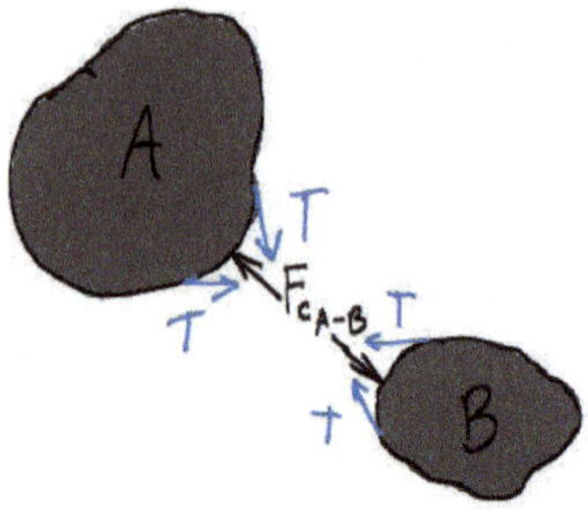

Intergranular contact force due to surface tension.

The surface tension of water explains why the water does not drain out of a wet sand castle or a moist ball of clay. Negative water pressures make the water stick to the particles and pull the particles to each other, friction at the particle contacts make a sand castle stable. But as soon as a wet sand castle is submerged below a free water surface, the negative pressures are lost and the castle collapses. Considering the effective stress equation, $\sigma' = \sigma - u$, if the water pressure is negative, the effective stress may be positive, even on a free surface (a surface where the total normal stress is zero). The negative pore pressure pulls the particles together and causes compressive particle to particle contact forces.

Negative pore pressures in clayey soil can be much more powerful than those in sand. Negative pore pressures explain why clay soils shrink when they dry and swell as they are wetted. The swelling and shrinkage can cause major distress, especially to light structures and roads.

Shrinkage caused by drying

Later sections of this article address the pore water pressures for seepage and consolidation problems.

SOIL CLASSIFICATION

Geotechnical engineers classify the soil particle types by performing tests on disturbed (dried, passed through sieves, and remolded) samples of the soil. This provides information about the characteristics of the soil grains themselves. It should be noted that classification of the types of grains present in a soil does not account for important effects of the structure or fabric of the soil, terms that describe compactness of the particles and patterns in the arrangement of particles in a load carrying framework as well as the pore size and pore fluid distributions. Engineering geologists also classify soils based on their genesis and depositional history.

Classification of Soil Grains

In the US and other countries, the Unified Soil Classification System (USCS) is often used for soil classification. Other classification systems include the British Standard BS5390 and the AASHTO soil classification system.

Classification of Sands and Gravels

In the USCS, gravels (given the symbol G) and sands (given the symbol S) are classified according to their grain size distribution. For the USCS, gravels may be given the classification symbol GW (well-graded gravel), GP (poorly graded gravel), GM (gravel with a large amount of silt), or GC (gravel with a large amount of clay). Likewise sands may be classified as being SW, SP, SM or SC. Sands and gravels with a small but non-negligible amount of fines (5–12%) may be given a dual classification such as SW-SC.

Atterberg Limits

Clays and Silts, often called 'fine-grained soils', are classified according to their Atterberg limits; the most commonly used Atterberg limits are the **Liquid Limit** (denoted by LL or w_l), **Plastic Limit** (denoted by PL or w_p), and **Shrinkage Limit** (denoted by SL).

The Liquid Limit is the water content at which the soil behavior transitions from that of a liquid to that of a plastic solid. The Plastic Limit is the water content at which the soil behavior transitions from that of a plastic solid to a brittle solid. The Shrinkage Limit corresponds to a water content below which the soil will not shrink as it dries.

As the transitions from one state to another are gradual, the tests have adopted arbitrary definitions to determine the boundaries of the states. The liquid limit is determined by measuring the water content for which a groove closes after 25 blows in a standard test. Alternatively, a fall cone test apparatus may be use to measure the liquid limit. The undrained shear strength of remolded soil at the liquid limit is approximately 2 kPa. The Plastic Limit is the water content below which it is not possible to roll by hand the soil into 3 mm diameter cylinders. The soil cracks or breaks up as it is rolled down to this diameter. Remolded soil at the plastic limit is quite stiff, having an undrained shear strength of the order of about 200 kPa. The **Plasticity Index** of a particular soil specimen is defined as the difference between the Liquid Limit and the Plastic Limit of the specimen; it is an indicator of how much water the soil particles in the specimen can absorb, and correlates with many engineering properties.

Classification of Silts and Clays

According to the Unified Soil Classification System (USCS), silts and clays are classified by plotting the values of their plasticity index and liquid limit on a plasticity chart. The A-Line on the chart separates clays (given the USCS symbol C) from silts (given the symbol M). LL=50% separates high plasticity soils (given the modifier symbol H) from low plasticity soils (given the modifier symbol L). A soil that plots above the A-line and has LL>50% would, for example, be classified as CH. Other possible classifications of silts and clays are ML, CL and MH. If the Atterberg limits plot in the"hatched" region on the graph near the origin, the soils are given the dual classification 'CL-ML'.

Indices Related to soil strength

Liquidity Index

The effects of the water content on the strength of saturated remolded soils can be quantified by the use of the liquidity index, LI:

$$LI = \frac{w - PL}{LL - PL}$$

When the LI is 1, remolded soil is at the liquid limit and it has an undrained shear strength of about 2 kPa. When the soil is at the plastic limit, the LI is 0 and the undrained shear strength is about 200 kPa.

Relative Density

The density of sands (cohesionless soils) is often characterized by the relative density, D_r

$$D_r = \frac{e_{max} - e}{e_{max} - e_{min}} 100\%$$

where: e_{max} is the "maximum void ratio" corresponding to a very loose state, e_{min} is the "minimum void ratio" corresponding to a very dense state and e is the in situ void ratio. Methods used to calculate relative density are defined in ASTM D4254-00(2006).

Thus if $D_r = 100\%$ the sand or gravel is very dense, and if $D_r = 0\%$ the soil is extremely loose and unstable.

SEEPAGE: STEADY STATE FLOW OF WATER

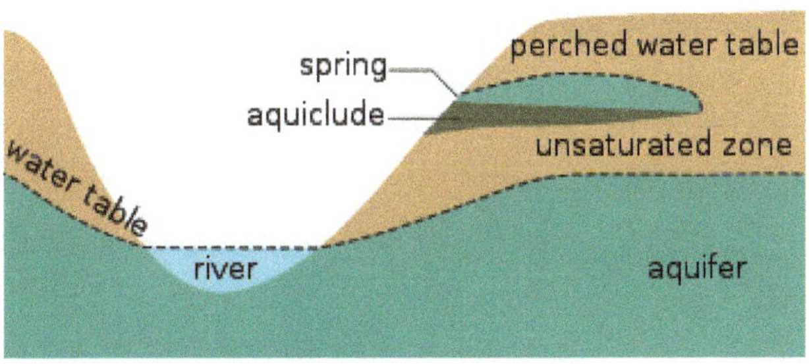

A cross section showing the water table varying with surface topography as well as a perched water table.

If fluid pressures in a soil deposit are uniformly increasing with depth according to $u = \rho_w g z_w$ then hydrostatic conditions will prevail and the fluids will not be flowing through the soil. z_w is the depth below the water table. However, if the water table is sloping or there is a perched water table as indicated in the accompanying sketch, then **seepage** will occur. For steady state seepage, the seepage velocities are not varying with time. If the water tables are changing levels with time, or if the soil is in the process of consolidation, then steady state conditions do not apply.

Darcy›s law

Darcy's law states that the volume of flow of the pore fluid through a porous medium per unit time is proportional to the rate of change of excess fluid pressure with distance. The constant of proportionality includes the viscosity of the fluid and the intrinsic permeability of the soil. For the simple case of a horizontal tube filled with soil

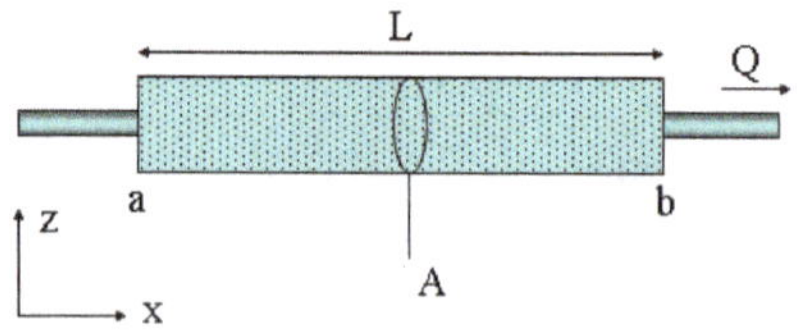

Diagram showing definitions and directions for Darcy's law.

$$Q = \frac{-KA}{\mu}\frac{(u_b - u_a)}{L}$$

The total discharge, Q (having units of volume per time, e.g., ft^3/s or m^3/s), is proportional to the intrinsic permeability, K, the cross sectional area, A, and rate of pore pressure change with distance, $\frac{u_b - u_a}{L}$, and inversely proportional to the dynamic viscosity of the fluid, μ. The negative sign is needed because fluids flow from high pressure to low pressure. So if the change in pressure is negative (in the x-direction) then the flow will be positive (in the x-direction). The above equation works well for a horizontal tube, but if the tube was inclined so that point b was a different elevation than point a, the equation would not work. The effect of elevation is accounted for by replacing the pore pressure by excess pore pressure, u_e defined as:

$$u_e = u - \rho_w g z$$

where z is the depth measured from an arbitrary elevation reference (datum). Replacing u by u_e we obtain a more general equation for flow:

$$Q = \frac{-KA}{\mu}\frac{(u_{e,b} - u_{e,a})}{L}$$

Dividing both sides of the equation by A, and expressing the rate of change of excess pore pressure as a derivative, we obtain a more general equation for the apparent velocity in the x-direction:

$$v_x = \frac{-K}{\mu}\frac{du_e}{dx}$$

where $v_x = Q/A$ has units of velocity and is called the Darcy velocity (or the specific discharge, filtration velocity, or superficial velocity). The pore or interstitial velocity v_{px} is the average velocity of fluid molecules in the pores; it is related to the Darcy velocity and the porosity n through the Dupuit-Forchheimer relationship

$$v_{px} = \frac{v_x}{n}$$

(Some authors use the term seepage velocity to mean the Darcy velocity, while others use it to mean the pore velocity.)

Civil engineers predominantly work on problems that involve water and predominantly work on problems on earth (in earth›s gravity). For this class of problems, civil engineers will often write Darcy›s law in a much simpler form:

$$v = ki$$

where k is the hydraulic conductivity, defined as $k = \frac{K\rho_w g}{\mu_w}$, and i is the hydraulic gradient. The hydraulic gradient is the rate of change of total head with distance. The total head, h at a point is defined as the height (measured relative to the datum) to which water would rise in a piezometer at that point. The total head is related to the excess water pressure by:

$$u_e = \rho_w gh + Constant$$

and the $Constant$ is zero if the datum for head measurement is chosen at the same elevation as the origin for the depth, z used to calculate u_e.

Typical values of Hydraulic Conductivity

Values of hydraulic conductivity, k, can vary by many orders of magnitude depending on the soil type. Clays may have hydraulic conductivity as small as about $10^{-12}\frac{m}{s}$, gravels may have hydraulic conductivity up to about $10^{-1}\frac{m}{s}$. Layering and heterogeneity and disturbance during the sampling and testing process make the accurate measurement of soil hydraulic conductivity a very difficult problem.

Flownets

Darcy's Law applies in one, two or three dimensions. In two or three dimensions, steady state seepage is described by Laplace's equation. Computer programs are available to solve this equation. But traditionally two-dimensional seepage problems were solved using a graphical procedure known as flownet. One set of lines in the flownet are in the direction of the water flow (flow lines), and the other set of lines are in the direction of constant total head (equipotential lines). Flownets may be used to estimate the quantity of seepage under dams and sheet piling.

A plan flow net to estimate flow of water from a stream to a discharging well

Seepage forces and erosion

When the seepage velocity is great enough, erosion can occur because of the frictional drag exerted on the soil particles. Vertically upwards seepage is a source of danger on the downstream side of sheet piling and beneath the toe of a dam or levee. Erosion of the soil, known as "soil piping", can lead to failure of the structure and to sinkhole formation. Seeping water removes soil, starting from the exit point of the seepage, and erosion advances upgradient. The term "sand boil" is used to describe the appearance of the discharging end of an active soil pipe.

Seepage Pressures

Seepage in an upward direction reduces the effective stress within the soil. When the water pressure at a point in the soil is equal to the total vertical stress at that point, the effective stress is zero and the soil has no frictional resistance to deformation. For a surface layer, the vertical effective stress becomes zero within the layer when the upward hydraulic gradient is equal to the critical gradient. At zero effective stress soil has very little strength and layers of relatively impermeable soil may heave up due to the underlying water pressures. The loss in strength due to upward seepage is a common contributor to levee failures. The condition of zero effective stress associated with upward seepage is also called liquefaction, quicksand, or a boiling condition. Quicksand was so named because the soil particles move around and appear to be 'alive' (the biblical meaning of 'quick' – as opposed to 'dead'). (Note that it is not possible to be 'sucked down' into quicksand. On the contrary, you would float with about half your body out of the water.)

CONSOLIDATION: TRANSIENT FLOW OF WATER

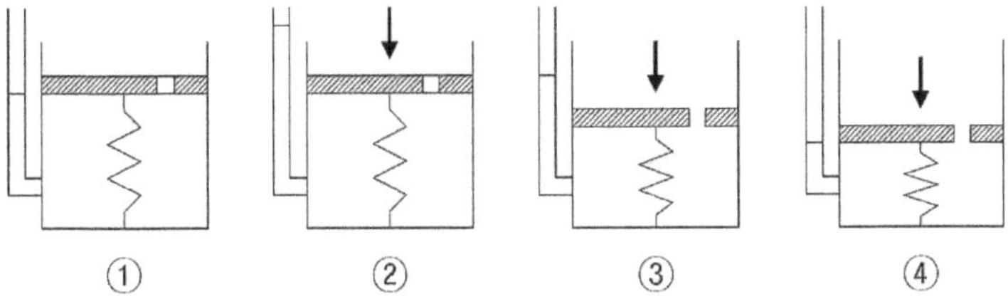

Consolidation analogy. The piston is supported by water underneath and a spring. When a load is applied to the piston, water pressure increases to support the load. As the water slowly leaks through the small hole, the load is transferred from the water pressure to the spring force.

Consolidation is a process by which soils decrease in volume. It occurs when stress is applied to a soil that causes the soil particles to pack together more tightly, therefore reducing volume. When this occurs in a soil that is saturated with water, water will be squeezed out of the soil. The time required to squeeze the water out of a thick deposit of clayey soil layer might be years. For a layer of sand, the water may be squeezed out in a matter of seconds. A building foundation or construction of a new embankment will cause the soil below to consolidate and this will cause settlement which in turn may cause distress to the building or embankment. Karl Terzaghi developed the theory of consolidation which enables prediction of the amount of settlement and the time required for the settlement to occur. Soils are tested with an oedometer test to determine their compression index and coefficient of consolidation.

When stress is removed from a consolidated soil, the soil will rebound, drawing water back into the pores and regaining some of the volume it had lost in the consolidation process. If the stress is reapplied, the soil will re-consolidate again along a recompression curve, defined by the recompression index. Soil that has been consolidated to a large pressure and has been subsequently unloaded is considered to be overconsolidated. The maximum past vertical effective stress is termed the preconsolidation stress. A soil which is currently experiencing the maximum past vertical effective stress is said to be normally consolidated. The overconsolidation ratio, (OCR) is the ratio of the maximum past vertical effective stress to the current vertical effective stress. The OCR is significant for two reasons: firstly, because the compressibility of normally consolidated soil is significantly larger than that for overconsolidated soil, and secondly, the shear behavior and dilatancy of clayey soil are related to the OCR through critical state soil mechanics; highly overconsolidated clayey soils are dilatant, while normally consolidated soils tend to be contractive.

SHEAR BEHAVIOR: STIFFNESS AND STRENGTH

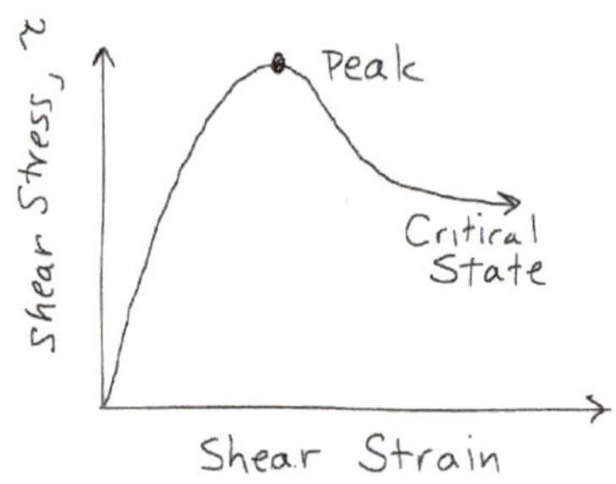

Typical stress strain curve for a drained dilatant soil.

The shear strength and stiffness of soil determines whether or not soil will be stable or how much it will deform. Knowledge of the strength is necessary to determine if a slope will be stable, if a building or bridge might settle too far into the ground, and the limiting pressures on a retaining wall. It is important to distinguish between failure of a soil element and the failure of a geotechnical structure (e.g., a building foundation, slope or retaining wall); some soil elements may reach their peak strength prior to failure of the structure. Different criteria can be used to define the "shear strength" and the "yield point" for a soil element from a stress–strain curve. One may define the peak shear strength as the peak of a stress–strain curve, or the shear strength at critical state as the value after large strains when the shear resistance levels off. If the stress–strain curve does not stabilize before the end of shear strength test, the "strength" is sometimes considered to be the shear resistance at 15–20% strain. The shear strength of soil depends on many factors including the effective stress and the void ratio.

The shear stiffness is important, for example, for evaluation of the magnitude of deformations of foundations and slopes prior to failure and because it is related to the shear wave velocity. The slope of the initial, nearly linear, portion of a plot of shear stress as a function of shear strain is called the shear modulus

Friction, Interlocking and Dilation

Angle of repose.

Soil is an assemblage of particles that have little to no cementation while rock (such as sandstone) may consist of an assembly of particles that are strongly cemented together by chemical bonds. The shear strength of soil is primarily due to interparticle friction and therefore, the shear resistance on a plane is approximately proportional to the effective normal stress on that plane. The angle of internal friction is thus closely related to the maximum stable slope angle, often called the angle of repose.

But in addition to friction, soil derives significant shear resistance from interlocking of grains. If the grains are densely packed, the grains tend to spread apart from each other as they are subject to shear strain. The expansion of the particle matrix due to shearing was called dilatancy by Osborne Reynolds. If one considers the energy required to shear an assembly of particles there is energy input by the shear force, T, moving a distance, x and there is also energy input by the normal force, N, as the sample expands a distance, y. Due to the extra energy required for the particles to dilate against the confining pressures, dilatant soils have a greater peak strength than contractive soils. Furthermore, as dilative soil grains dilate, they become looser (their void ratio increases), and their rate of dilation decreases until they reach a critical void ratio. Contractive soils become denser as they shear, and their rate of contraction decreases until they reach a critical void ratio.

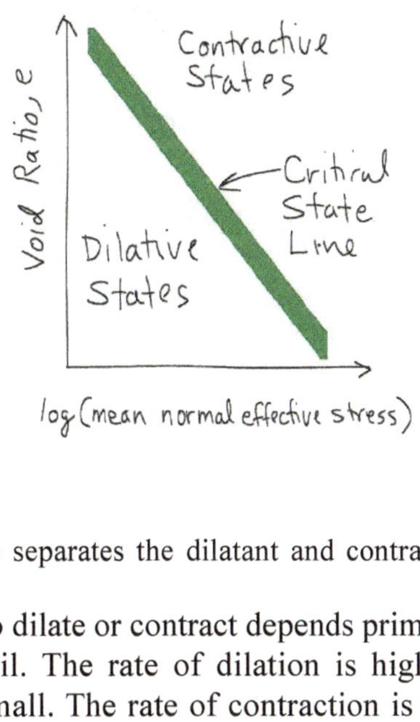

A critical state line separates the dilatant and contractive states for soil.

The tendency for a soil to dilate or contract depends primarily on the confining pressure and the void ratio of the soil. The rate of dilation is high if the confining pressure is small and the void ratio is small. The rate of contraction is high if the confining pressure is large and the void ratio is large. As a first approximation, the regions of contraction and dilation are separated by the critical state line.

Failure Criteria

After a soil reaches the critical state, it is no longer contracting or dilating and the shear stress on the failure plane τ_{crit} is determined by the effective normal stress on the failure plane σ'_n and critical state friction angle ϕ'_{crit}:

$$\tau_{crit} = \sigma'_n \tan \phi'_{crit}$$

The peak strength of the soil may be greater, however, due to the interlocking (dilatancy) contribution. This may be stated:

$$\tau_{peak} = \sigma'_n \tan \phi'_{peak}$$

Where $\phi'_{peak} > \phi'_{crit}$. However, use of a friction angle greater than the critical state value for design requires care. The peak strength will not be mobilized everywhere at the same time in a practical problem such as a foundation, slope or retaining wall. The critical state friction angle is not nearly as variable as the peak friction angle and hence it can be relied upon with confidence.

Not recognizing the significance of dilatancy, Coulomb proposed that the shear strength of soil may be expressed as a combination of adhesion and friction components:

$$\tau_f = c' + \sigma'_f \tan \phi'$$

It is now known that the c' and ϕ' parameters in the last equation are not fundamental soil properties. In particular, c' and ϕ' are different depending on the magnitude of effective stress. According to Schofield (2006), the longstanding use of c' in practice has led many engineers to wrongly believe that c' is a fundamental parameter. This assumption that c' and ϕ' are constant can lead to overestimation of peak strengths.

Structure, Fabric, and Chemistry

In addition to the friction and interlocking (dilatancy) components of strength, the structure and fabric also play a significant role in the soil behavior. The structure and fabric include factors such as the spacing and arrangement of the solid particles or the amount and spatial distribution of pore water; in some cases cementitious material accumulates at particle-particle contacts. Mechanical behavior of soil is affected by the density of the particles and their structure or arrangement of the particles as well as the amount and spatial distribution of fluids present (e.g., water and air voids). Other factors include the electrical charge of the particles, chemistry of pore water, chemical bonds (i.e. cementation -particles connected through a solid substance such as recrystallized calcium carbonate)

Drained and Undrained Shear

The presence of nearly incompressible fluids such as water in the pore spaces affects the ability for the pores to dilate or contract.

If the pores are saturated with water, water must be sucked into the dilating pore spaces to fill the expanding pores (this phenomenon is visible at the beach when apparently dry spots form around feet that press into the wet sand).

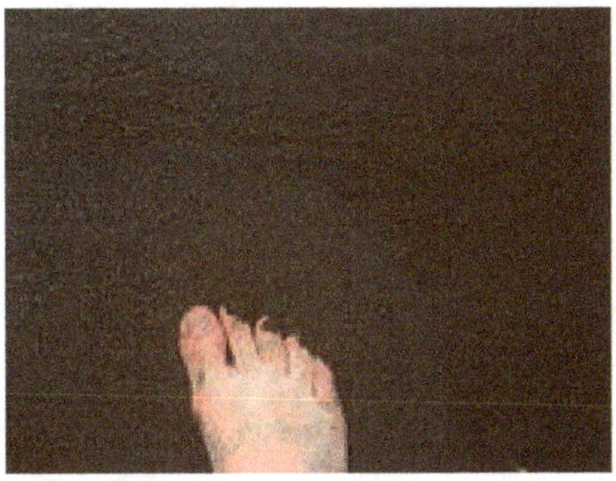

Foot pressing in soil causes soil to dilate, drawing water from the surface into the pores

Similarly, for contractive soil, water must be squeezed out of the pore spaces to allow contraction to take place.

Dilation of the voids causes negative water pressures that draw fluid into the pores, and contraction of the voids causes positive pore pressures to push the water out of the pores. If the rate of shearing is very large compared to the rate that water can be sucked into or squeezed out of the dilating or contracting pore spaces, then the shearing is called undrained shear, if the shearing is slow enough that the water pressures are negligible, the shearing is called drained shear. During undrained shear, the water pressure u changes depending on volume change tendencies. From the effective stress equation, the change in u directly effects the effective stress by the equation:

$$\sigma' = \sigma - u$$

and the strength is very sensitive to the effective stress. It follows then that the undrained shear strength of a soil may be smaller or larger than the drained shear strength depending upon whether the soil is contractive or dilative.

Shear Tests

Strength parameters can be measured in the laboratory using direct shear test, triaxial shear test, simple shear test, fall cone test and (hand) shear vane test; there are numerous other devices and variations on these devices used in practice today. Tests conducted to characterize the strength and stiffness of the soils in the ground include the Cone penetration test and the Standard penetration test.

Other Factors

The stress–strain relationship of soils, and therefore the shearing strength, is affected by:

- soil composition (basic soil material): mineralogy, grain size and grain size distribution, shape of particles, pore fluid type and content, ions on grain and in pore fluid.
- state (initial): Define by the initial void ratio, effective normal stress and shear stress (stress history). State can be describe by terms such as: loose, dense, overconsolidated, normally consolidated, stiff, soft, contractive, dilative, etc.

- structure: Refers to the arrangement of particles within the soil mass; the manner in which the particles are packed or distributed. Features such as layers, joints, fissures, slickensides, voids, pockets, cementation, etc., are part of the structure. Structure of soils is described by terms such as: undisturbed, disturbed, remolded, compacted, cemented; flocculent, honey-combed, single-grained; flocculated, deflocculated; stratified, layered, laminated; isotropic and anisotropic.
- Loading conditions: Effective stress path -drained, undrained, and type of loading -magnitude, rate (static, dynamic), and time history (monotonic, cyclic).

APPLICATIONS

Lateral earth pressure

Lateral earth stress theory is used to estimate the amount of stress soil can exert perpendicular to gravity. This is the stress exerted on retaining walls. A lateral earth stress coefficient, K, is defined as the ratio of lateral (horizontal) effective stress to vertical effective stress for cohesionless soils ($K=\sigma'_h/\sigma'_v$). There are three coefficients: at-rest, active, and passive. At-rest stress is the lateral stress in the ground before any disturbance takes place. The active stress state is reached when a wall moves away from the soil under the influence of lateral stress, and results from shear failure due to reduction of lateral stress. The passive stress state is reached when a wall is pushed into the soil far enough to cause shear failure within the mass due to increase of lateral stress. There are many theories for estimating lateral earth stress; some are empirically based, and some are analytically derived.

Bearing Capacity

The bearing capacity of soil is the average contact stress between a foundation and the soil which will cause shear failure in the soil. Allowable bearing stress is the bearing capacity divided by a factor of safety. Sometimes, on soft soil sites, large settlements may occur under loaded foundations without actual shear failure occurring; in such cases, the allowable bearing stress is determined with regard to the maximum allowable settlement. It is important during construction and design stage of a project to evaluate the subgrade strength. The California Bearing Ratio (CBR) test is commonly used to determine the suitability of a soil as a subgrade for design and construction. The field Plate Load Test is commonly used to predict the deformations and failure characteristics of the soil/subgrade and modulus of subgrade reaction (ks). The Modulus of subgrade reaction (ks) is used in foundation design, soil-structure interaction studies and design of highway pavements.

Slope Stability

The field of slope stability encompasses the analysis of static and dynamic stability of slopes of earth and rock-fill dams, slopes of other types of embankments, excavated slopes, and natural slopes in soil and soft rock.

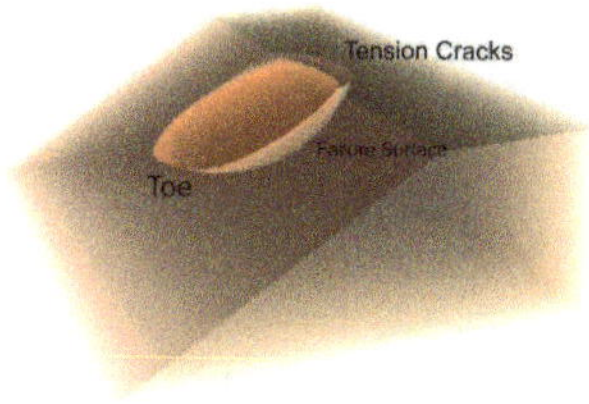

Simple slope slip section

As seen to the right, earthen slopes can develop a cut-spherical weakness zone. The probability of this happening can be calculated in advance using a simple 2-D circular analysis package... A primary difficulty with analysis is locating the most-probable slip plane for any given situation. Many landslides have been analyzed only after the fact.

RECENT DEVELOPMENTS

A recent finding in soil mechanics is that soil deformation can be described as the behavior of a dynamical system. This approach to soil mechanics is referred to as Dynamical Systems based Soil Mechanics (DSSM). DSSM holds simply that soil deformation is a Poisson process in which particles move to their final position at random shear strains.

The basis of DSSM is that soils (including sands) can be sheared till they reach a steady-state condition at which, under conditions of constant strain-rate, there is no change in shear stress, effective confining stress, and void ratio. The steady-state was formally defined by Steve J. Poulos an Associate Professor at the Soil Mechanics Department of Harvard University, who built off a hypothesis that Arthur Casagrande was formulating towards the end of his career. The steady state condition is not the same as the "critical state" condition.It differs from the critical state in that it specifies a statistically constant structure at the steady state. The steady-state values are also very slightly dependent on the strain-rate.

Many systems in nature reach steady-states and dynamical systems theory is used to describe such systems. Soil shear can also be described as a dynamical system. The physical basis of the soil shear dynamical system is a Poisson process in which particles move to the steady-state at random shear strains.Joseph generalized this—particles move to their final position (not just steady-state) at random shear-strains. Because of its origins in the steady state concept DSSM is sometimes informally called "Harvard soil mechanics."

DSSM provides for very close fits to stress–strain curves, including for sands. Because it tracks conditions on the failure plane, it also provides close fits for the post failure region of sensitive clays and silts something that other theories are not able to do. Additionally DSSM explains key relationships in soil mechanics that to date have simply been taken for granted, for example, why normalized undrained peak shear strengths vary with the log of the over consolidation ratio and why stress–strain curves normalize with the initial effective confining stress; and why in one-dimensional consolidation the void ratio must vary with the log of the effective vertical stress, why the end-of-primary curve is unique for static load increments, and why the ratio of the creep value Cα to the compression index Cc must be approximately constant for a wide range of soils.

REFERENCES

1. Mitchell, J.K., and Soga, K. (2005) Fundamentals of soil behavior, Third edition, John Wiley and Sons, Inc., ISBN 978-0-471-46302-3.
2. Santamarina, J.C., Klein, K.A., & Fam, M.A. (2001). Soils and Waves: Particulate Materials Behavior, Characterization and Process Monitoring. Wiley. ISBN 978-0-471-49058-6. .
3. Powrie, W., Spon Press, 2004, Soil Mechanics – 2nd ed ISBN 0-415-31156-X
4. A Guide to Soil Mechanics, Bolton, Malcolm,Macmillan Press, 1979. ISBN 0-333-18931-0
5. Fang, Y., Spon Press, 2006, Introductory Geotechnical Engineering
6. Lambe, T. William & Robert V. Whitman. Soil Mechanics. Wiley, 1991; p. 29. ISBN 978-0-471-51192-2
7. ASTM Standard Test Methods of Particle-Size Distribution (Gradation) of Soils using Sieve Analysis. http://www.astm.org/Standards/D6913.htm
8. Ozhovan, M.I.; Dmitriev, I.E.; Batyukhnova, O.G. (1993). "Fractal structure of pores in clay soil". Atomic Energy. **74** (3): 241–243. doi:10.1007/BF00739059.
9. Holtz, R.D, and Kovacs, W.D., 1981. An Introduction to Geotechnical Engineering. Prentice-Hall, Inc. page 448
10. Classification of Soils for Engineering Purposes: Annual Book of ASTM Standards". D 2487-83. 04.08. American Society for Testing and Materials. 1985: 395–408.
11. Wood, David Muir, Soil Behavior and Critical State Soil Mechanics, Cambridge University Press, 1990, ISBN 0-521-33249-4
12. Disturbed soil properties and geotechnical design, Schofield, Andrew N.,Thomas Telford, 2006. ISBN 0-7277-2982-9
13. ASTM Standard Test Methods for Minimum Index Density and Unit Weight of Soils and Calculation of Relative Density. http://www.astm.org/Standards/D4254.htm
14. Smith, I. (2013) Smith's elements of soil mechanics, 8th edition, John Wiley and Sons, Inc., ISBN 978-1-405-13370-8.
15. Delleur, Jacques W. (2007) The handbook of groundwater engineering, Taylor & Francis, ISBN 978-0-849-34316-2.
16. Cedergren, Harry R. (1977), Seepage, Drainage, and Flow Nets, Wiley. ISBN 0-471-14179-8
17. Jones, J. A. A. (1976). "Soil piping and stream channel initiation". Water

Resources Research. **7** (3): 602–610. Bibcode:1971WRR.....7..602J. doi:10.1029/WR007i003p00602.

18. Dooley, Alan (June 2006). "Sandboils 101: Corps has experience dealing with common flood danger". Engineer Update. US Army Corps of Engineers. Archived from the original on 2006-07-27. Retrieved 2006-08-29.
19. Terzaghi, K., Peck, R.B., and Mesri, G. 1996. Soil Mechanics in Engineering Practice. Third Edition, John Wiley & Sons, Inc. Article 18, page 135.
20. Terzaghi, K., 1943, Theoretical Soil Mechanics, John Wiley and Sons, New York
21. Terzaghi, K., Peck, R.B., Mesri, G. (1996) Soil mechanics in Engineering Practice, Third Edition, John Wiley & Sons, Inc.,ISBN 0-471-08658-4
22. Poulos, S. J. 1989. Advance Dam Engineering for Design, Construction, and Rehabilitation: Liquefaction Related Phenomena. Ed. Jansen, R.B, Van Nostrand Reinhold, pages 292–297.
23. Eka Putri, E.; Kameswara Rao, N.S.V.; Mannan, M.A. (2012). "Evaluation of Modulus of Elasticity and Modulus of Subgrade Reaction of Soils Using CBR Test". Journal of Civil Engineering Research. **2** (1): 34–40. doi:10.5923/j.jce.20120201.05.
24. US Army Corps of Engineers Manual on Slope Stability
25. Slope Stability Calculator". Retrieved 2006-12-14.
26. Chugh, A.K. (2002). "A method for locating critical slip surfaces in slope stability analysis: Discussion". Canadian Geotechnical Journal. **39** (3): 765–770. doi:10.1139/t02-042.
27. Poulos, Steve J. (1981). "The Steady State of Deformation". Journal of Geotechnical Engineering. **107** (GT5): 553–562.
28. Joseph, Paul G. (2009). "Constitutive Model of Soil Based on a Dynamical Systems Approach". Journal of Geotechnical and Geoenvironmental Engineering. **135** (8): 1155–1158. doi:10.1061/(asce)gt.1943-5606.0000001.
29. Joseph, Paul G. (2010). "A Dynamical Systems Based Approach to Soil Shear". Geotechnique. **LX** (10): 807–812. doi:10.1680/geot.9.p.001.
30. Joseph, Paul G. (2012). "Physical Basis and Validation of a Constitutive Model for Soil Shear Derived from Micro-Structural Changes". International Journal of Geomechanics. **13** (4): 365–383. doi:10.1061/(asce)gm.1943-5622.0000209.
31. Joseph, Paul G. (2014). "Generalized soil deformation model based on dynamical systems theory". Geotechnical Research. **1** (1): 32–42. doi:10.1680/geores.14.00004.
32. Joseph, Paul G. (2013). "Dynamical systems based soil mechanics (DSSM), a short self-study course".

CHAPTER 3

Soil Forming and Degradation Processes

Chapter Contents

The formation of a new soil depends on the presence of new soil material – obtained either by denudation - the scraping clean of a surface by action of wind, water or ice - or deposition of new materials from erosion elsewhere - for example river gravels, rock-falls or blowing sand - or the formation of new rocks by uplift of the ocean bed or volcanic action.

SOIL-FORMING PROCESSES

Soil formation, or pedogenesis, is the combined effect of physical, chemical, biological and anthropogenic processes working on soil parent material. Soil is said to be formed when organic matter has accumulated and colloids are washed downward, leaving deposits of clay, humus, iron oxide, carbonate, and gypsum, producing a distinct layer called the B horizon. This is a somewhat arbitrary definition as mixtures of sand, silt, clay and humus will support biological and agricultural activity before that time. These constituents are moved from one level to another by water and animal activity. As a result, layers (horizons) form in the soil profile. The alteration and movement of materials within a soil causes the formation of distinctive soil horizons.

How soil formation proceeds is influenced by at least five classic factors that are intertwined in the evolution of a soil. They are: parent material, climate, topography (relief), organisms, and time. When reordered to climate, relief, organisms, parent material, and time, they form the acronym CROPT.

An example of the development of a soil would begin with the weathering of lava flow bedrock, which would produce the purely mineral-based parent material from which the soil texture forms. Soil development would proceed most rapidly from bare rock of recent flows in a warm climate, under heavy and frequent rainfall. Under such conditions, plants become established very quickly on basaltic lava, even though there is very little organic material. The plants are supported by the porous rock as it is filled with nutrient-bearing water that carries dissolved minerals from the rocks and guano. Crevasses and pockets, local topography of the rocks, would hold fine materials and harbour plant roots. The developing plant roots are associated with mycorrhizal fungi that assist in breaking up the porous lava, and by these means organic matter and a finer mineral soil accumulate with time.

Parent Material

The mineral material from which a soil forms is called parent material. Rock, whether its origin is igneous, sedimentary, or metamorphic, is the source of all soil mineral materials and the origin of all plant nutrients with the exceptions of nitrogen, hydrogen and carbon. As the parent material is chemically and physically weathered, transported, deposited and precipitated, it is transformed into a soil.

Typical soil parent mineral materials are:

- Quartz: SiO_2
- Calcite: $CaCO_3$
- Feldspar: $KAlSi_3O_8$
- Mica (biotite): $K(Mg,Fe)_3AlSi_3O_{10}(OH)_2$

Classification of Parent Material

Loess field in Germany.

Parent materials are classified according to how they came to be deposited. Residual materials are mineral materials that have weathered in place from primary bedrock. Transported materials are those that have been deposited by water, wind, ice or gravity. Cumulose material is organic matter that has grown and accumulates in place.

Residual soils are soils that develop from their underlying parent rocks and have the same general chemistry as those rocks. The soils found on mesas, plateaux, and plains are residual soils. In the United States as little as three percent of the soils are residual.

Most soils derive from transported materials that have been moved many miles by wind, water, ice and gravity.

- Aeolian processes (movement by wind) are capable of moving silt and fine sand many hundreds of miles, forming loess soils (60–90 percent silt), common in the Midwest of North America and in Central Asia. Clay is seldom moved by wind as it forms stable aggregates.
- Water-transported materials are classed as either alluvial, lacustrine, or marine. Alluvial materials are those moved and deposited by flowing water. Sedimentary deposits settled in lakes are called lacustrine. Lake Bonneville and many soils around the Great Lakes of the United States are examples. Marine deposits, such as soils along the Atlantic and Gulf Coasts and in the Imperial Valley of California of the United States, are the beds of ancient seas that have been revealed as the land uplifted.
- Ice moves parent material and makes deposits in the form of terminal and lateral moraines in the case of stationary glaciers. Retreating glaciers leave smoother ground moraines and in all cases, outwash plains are left as alluvial deposits are moved downstream from the glacier.
- Parent material moved by gravity is obvious at the base of steep slopes as talus cones and is called colluvial material.

Cumulose parent material is not moved but originates from deposited organic material. This includes peat and muck soils and results from preservation of plant residues by the low oxygen content of a high water table. While peat may form sterile soils, muck soils may be very fertile.

Weathering of Parent Material

The weathering of parent material takes the form of physical weathering (disintegration), chemical weathering (decomposition) and chemical transformation. Generally, minerals that are formed under the high temperatures and pressures at great depths within the Earth's mantle are less resistant to weathering, while minerals formed at low temperature and pressure environment of the surface are more resistant to weathering. Weathering is usually confined to the top few meters of geologic material, because physical, chemical, and biological stresses generally decrease with depth. Physical disintegration begins as rocks that have solidified deep in the Earth are exposed to lower pressure near the surface and swell and become mechanically unstable. Chemical decomposition is a function of mineral solubility, the rate of which doubles with each 10 °C rise in temperature, but is strongly dependent on water to effect chemical changes. Rocks that will decompose in a few years in tropical climates will remain unaltered for millennia in deserts. Structural changes are the result of hydration, oxidation, and reduction.

- **Physical disintegration** is the first stage in the transformation of parent material into soil. Temperature fluctuations cause expansion and contraction of the rock, splitting it along lines of weakness. Water may then enter the cracks and freeze and cause the physical splitting of material along a path toward the center of the rock, while temperature gradients within the rock can cause exfoliation of "shells". Cycles of wetting and drying cause soil particles to be abraded to a finer size, as does the physical rubbing of material as it is moved by wind, water, and gravity. Water can deposit within rocks minerals that expand upon drying, thereby stressing the rock. Finally, organisms reduce parent material in size through the action of plant roots or digging on the part of animals.
- **Chemical decomposition** and **structural changes** result when minerals are made soluble by water or are changed in structure. The first three of the following list are solubility changes and the last three are structural changes.
- The **solution** of salts in water results from the action of bipolar water on ionic salt compounds producing a solution of ions and water, removing those minerals and reducing the rock's integrity.
- **Hydrolysis** is the transformation of minerals into polar molecules by the splitting of intervening water. This results in soluble acid-base pairs. For example, the hydrolysis of orthoclase-feldspar transforms it to acid silicate clay and basic potassium hydroxide, both of which are more soluble.
- In **carbonation**, the solution of carbon dioxide in water forms carbonic acid. Carbonic acid will transform calcite into more soluble calcium bicarbonate.
- **Hydration** is the inclusion of water in a mineral structure, causing it to swell and leaving it more stressed and easily decomposed.
- **Oxidation** of a mineral compound is the inclusion of oxygen in a mineral, causing it to increase its oxidation number and swell due to the relatively large size of oxygen, leaving it stressed and more easily attacked by water (hydrolysis) or carbonic acid (carbonation).
- **Reduction** the opposite of oxidation, means the removal of oxygen, hence the oxidation number of some part of the mineral is reduced, which occurs when oxygen is scarce. The reduction of minerals leaves them electrically unstable, more soluble and internally stressed and easily decomposed.

Of the above, hydrolysis and carbonation are the most effective.

Saprolite is a particular example of a residual soil formed from the transformation of granite, metamorphic and other types of bedrock into clay minerals. Often called "weathered granite", saprolite is the result of weathering processes that include: hydrolysis, chelation from organic compounds, hydration (the solution of minerals in water with resulting cation and anion pairs) and physical processes that include freezing and thawing. The mineralogical and chemical composition of the primary bedrock material, its physical features, including grain size and degree of consolidation, and the rate and type of weathering transforms the parent material into a different mineral. The texture, pH and mineral constituents of saprolite are inherited from its parent material.

Climate

The principal climatic variables influencing soil formation are effective precipitation (i.e., precipitation minus evapotranspiration) and temperature, both of which affect the rates of chemical, physical, and biological processes. The temperature and moisture both influence the organic matter content of soil through their effects on the balance between plant growth and microbial decomposition. Climate is the dominant factor in soil formation, and soils show the distinctive characteristics of the climate zones in which they form. For every 10 °C rise in temperature, the rates of biochemical reactions more than double. Mineral precipitation and temperature are the primary climatic influences on soil formation. If warm temperatures and abundant water are present in the profile at the same time, the processes of weathering, leaching, and plant growth will be maximized. Humid climates favor the growth of trees. In contrast, grasses are the dominant native vegetation in subhumid and semiarid regions, while shrubs and brush of various kinds dominate in arid areas.

Water is essential for all the major chemical weathering reactions. To be effective in soil formation, water must penetrate the regolith. The seasonal rainfall distribution, evaporative losses, site topography, and soil permeability interact to determine how effectively precipitation can influence soil formation. The greater the depth of water penetration, the greater the depth of weathering of the soil and its development. Surplus water percolating through the soil profile transports soluble and suspended materials from the upper to the lower layers. It may also carry away soluble materials in the surface drainage waters. Thus, percolating water stimulates weathering reactions and helps differentiate soil horizons. Likewise, a deficiency of water is a major factor in determining the characteristics of soils of dry regions. Soluble salts are not leached from these soils, and in some cases they build up to levels that curtail plant growth. Soil profiles in arid and semi-arid regions are also apt to accumulate carbonates and certain types of expansive clays.

The direct influences of climate include:

- A shallow accumulation of lime in low rainfall areas as caliche
- Formation of acid soils in humid areas
- Erosion of soils on steep hillsides
- Deposition of eroded materials downstream
- Very intense chemical weathering, leaching, and erosion in warm and humid regions where soil does not freeze

Climate directly affects the rate of weathering and leaching. Wind moves sand and smaller particles, especially in arid regions where there is little plant cover. The type and amount of precipitation influence soil formation by affecting the movement of ions and particles through the soil, and aid in the development of different soil profiles. Soil profiles are more distinct in wet and cool climates, where organic materials may accumulate, than in wet and warm climates, where organic materials are rapidly consumed.

The effectiveness of water in weathering parent rock material depends on seasonal and daily temperature fluctuations. Cycles of freezing and thawing constitute an effective mechanism which breaks up rocks and other consolidated materials.

Climate also indirectly influences soil formation through the effects of vegetation cover and biological activity, which modify the rates of chemical reactions in the soil.

Topography

The topography, or relief, is characterized by the inclination (slope), elevation, and orientation of the terrain. Topography determines the rate of precipitation or runoff and rate of formation or erosion of the surface soil profile. The topographical setting may either hasten or retard the work of climatic forces.

Steep slopes encourage rapid soil loss by erosion and allow less rainfall to enter the soil before running off and hence, little mineral deposition in lower profiles. In semiarid regions, the lower effective rainfall on steeper slopes also results in less complete vegetative cover, so there is less plant contribution to soil formation. For all of these reasons, steep slopes prevent the formation of soil from getting very far ahead of soil destruction. Therefore, soils on steep terrain tend to have rather shallow, poorly developed profiles in comparison to soils on nearby, more level sites.

In swales and depressions where runoff water tends to concentrate, the regolith is usually more deeply weathered and soil profile development is more advanced. However, in the lowest landscape positions, water may saturate the regolith to such a degree that drainage and aeration are restricted. Here, the weathering of some minerals and the decomposition of organic matter are retarded, while the loss of iron and manganese is accelerated. In such low-lying topography, special profile features characteristic of wetland soils may develop. Depressions allow the accumulation of water, minerals and organic matter and in the extreme, the resulting soils will be saline marshes or peat bogs. Intermediate topography affords the best conditions for the formation of an agriculturally productive soil.

Organisms

Soil is the most abundant ecosystem on Earth, but the vast majority of organisms in soil are microbes, a great many of which have not been described. There may be a population limit of around one billion cells per gram of soil, but estimates of the number of species vary widely from 50,000 per gram to over a million per gram of soil. The total number of organisms and species can vary widely according to soil type, location, and depth.

Plants, animals, fungi, bacteria and humans affect soil formation (see soil biomantle and stonelayer). Animals, soil mesofauna and micro-organisms mix soils as they form burrows and pores, allowing moisture and gases to move about. In the same way, plant roots open channels in soils. Plants with deep taproots can penetrate many metres through the different soil layers to bring up nutrients from deeper in the profile. Plants with fibrous roots that spread out near the soil surface have roots that are easily decomposed, adding organic matter. Micro-organisms, including fungi and bacteria, effect chemical exchanges between roots and soil and act as a reserve of nutrients.

Humans impact soil formation by removing vegetation cover with erosion as the result. Their tillage also mixes the different soil layers, restarting the soil formation process as less weathered material is mixed with the more developed upper layers.

Earthworms, ants and termites mix the soil as they burrow, significantly affecting soil formation. Earthworms ingest soil particles and organic residues, enhancing the availability of plant nutrients in the material that passes through their bodies. They aerate and stir the soil and increase the stability of soil aggregates, thereby assuring ready infiltration of water. In addition, as ants and termites build mounds, they transport soil materials from one horizon to another.

In general, the mixing of the soil by the activities of animals, sometimes called pedoturbation, tends to undo or counteract the tendency of other soil-forming processes that create distinct horizons. Termites and ants may also retard soil profile development by denuding large areas of soil around their nests, leading to increased loss of soil by erosion. Large animals such as gophers, moles, and prairie dogs bore into the lower soil horizons, bringing materials to the surface. Their tunnels are often open to the surface, encouraging the movement of water and air into the subsurface layers. In localized areas, they enhance mixing of the lower and upper horizons by creating, and later refilling, underground tunnels. Old animal burrows in the lower horizons often become filled with soil material from the overlying A horizon, creating profile features known as crotovinas.

Vegetation impacts soils in numerous ways. It can prevent erosion caused by excessive rain that might result from surface runoff. Plants shade soils, keeping them cooler and slow evaporation of soil moisture, or conversely, by way of transpiration, plants can cause soils to lose moisture. Plants can form new chemicals that can break down minerals and improve the soil structure. The type and amount of vegetation depends on climate, topography, soil characteristics, and biological factors. Soil factors such as density, depth, chemistry, pH, temperature and moisture greatly affect the type of plants that can grow in a given location. Dead plants and fallen leaves and stems begin their decomposition on the surface. There, organisms feed on them and mix the organic material with the upper soil layers; these added organic compounds become part of the soil formation process.

Human activities widely influence soil formation. For example, it is believed that Native Americans regularly set fires to maintain several large areas of prairie grasslands in Indiana and Michigan. In more recent times, human destruction of natural vegetation and subsequent tillage of the soil for crop production has abruptly modified soil formation. Likewise, irrigating an arid region of soil drastically influences the soil-forming factors, as does adding fertilizer and lime to soils of low fertility.

Time

Time is a factor in the interactions of all the above. While a mixture of sand, silt and clay constitute the texture of a soil and the aggregation of those components produces peds, the development of a distinct B horizon marks the development of a soil. With time, soils will evolve features that depend on the interplay of the prior listed soil-forming factors. It takes decades to several thousand years for a soil to develop a profile. That time period depends strongly on climate, parent material, relief, and biotic activity. For example, recently deposited material from a flood exhibits no soil development as there has not been enough time for the material to form a structure that further defines soil. The original soil surface is buried, and the formation process must begin anew for this deposit. Over time the soil will develop a profile that depends on the intensities of biota and climate. While a soil can achieve relative stability of its properties for extended periods, the soil life cycle ultimately ends in soil conditions that leave it vulnerable to erosion. Despite the inevitability of soil retrogression and degradation, most soil cycles are long.

Soil-forming factors continue to affect soils during their existence, even on "stable" landscapes that are long-enduring, some for millions of years. Materials are deposited on top or are blown or washed from the surface. With additions, removals and alterations, soils are always subject to new conditions. Whether these are slow or rapid changes depends on climate, topography and biological activity.

Land degradation refers to a human-induced or natural process which impairs the capacity of land to function. Soils degradation involves the acidification, contamination, desertification, erosion or salination.

Soil acidification is beneficial in the case of alkaline soils, but it degrades land when it lowers crop productivity and increases soil vulnerability to contamination and erosion. Soils are often initially acid because their parent materials were acid and initially low in the basic cations (calcium, magnesium, potassium and sodium). Acidification occurs when these elements are leached from the soil profile by rainfall or by the harvesting of forest or agricultural crops. Soil acidification is accelerated by the use of acid-forming nitrogenous fertilizers and by the effects of acid precipitation.

Soil contamination at low levels is often within a soil's capacity to treat and assimilate waste material. Soil biota can treat waste by transforming it; soil colloids can adsorb the waste material. Many waste treatment processes rely on this treatment capacity. Exceeding treatment capacity can damage soil biota and limit soil function. Derelict soils occur where industrial contamination or other development activity damages the soil to such a degree that the land cannot be used safely or productively. Remediation of derelict soil uses principles of geology, physics, chemistry and biology to degrade, attenuate, isolate or remove soil contaminants to restore soil functions and values. Techniques include leaching, air sparging, chemical amendments, phytoremediation, bioremediation and natural degradation.

Desertification is an environmental process of ecosystem degradation in arid and semi-arid regions, often caused by human activity. It is a common misconception that droughts cause desertification. Droughts are common in arid and semiarid lands. Well-managed lands can recover from drought when the rains return. Soil management tools include maintaining soil nutrient and organic matter levels, reduced tillage and increased cover. These practices help to control erosion and maintain productivity during periods when moisture is available. Continued land abuse during droughts, however, increases land degradation. Increased population and livestock pressure on marginal lands accelerates desertification. Erosion of soil is caused by water, wind, ice, and movement in response to gravity. More than one kind of erosion can occur simultaneously. Erosion is distinguished from weathering, since erosion also transports eroded soil away from its place of origin (soil in transit may be described as sediment). Erosion is an intrinsic natural process, but in many places it is greatly increased by human activity, especially poor land use practices. These include agricultural activities which leave the soil bare during times of heavy rain or strong winds, overgrazing, deforestation, and improper construction activity. Improved management can limit erosion. Soil conservation techniques which are employed include changes of land use (such as replacing erosion-prone crops with grass or other soil-binding plants), changes to the timing or type of agricultural operations, terrace building, use of erosion-suppressing cover materials (including cover crops and other plants), limiting disturbance during construction, and avoiding construction during erosion-prone periods.

A serious and long-running water erosion problem occurs in China, on the middle reaches of the Yellow River and the upper reaches of the Yangtze River. From the Yellow River, over 1.6 billion tons of sediment flow each year into the ocean. The sediment originates primarily from water erosion (gully erosion) in the Loess Plateau region of northwest China.

Soil piping is a particular form of soil erosion that occurs below the soil surface. It causes levee and dam failure, as well as sink hole formation. Turbulent flow removes soil starting at the mouth of the seep flow and the subsoil erosion advances up-gradient. The term sand boil is used to describe the appearance of the discharging end of an active soil pipe.

Desertification

Soil salination is the accumulation of free salts to such an extent that it leads to degradation of the agricultural value of soils and vegetation. Consequences include corrosion damage, reduced plant growth, erosion due to loss of plant cover and soil structure, and water quality problems due to sedimentation. Salination occurs due to a combination of natural and human-caused processes. Arid conditions favour salt accumulation. This is especially apparent when soil parent material is saline. Irrigation of arid lands is especially problematic. All irrigation water has some level of salinity. Irrigation, especially when it involves leakage from canals and overirrigation in the field, often raises the underlying water table. Rapid salination occurs when the land surface is within the capillary fringe of saline groundwater. Soil salinity control involves watertable control and flushing with higher levels of applied water in combination with tile drainage or another form of subsurface drainage.

Erosion control

RECLAMATION

Soils which contain high levels of particular clays, such as smectites, are often very fertile. For example, the smectite-rich clays of Thailand's Central Plains are among the most productive in the world.

Many farmers in tropical areas, however, struggle to retain organic matter in the soils they work. In recent years, for example, productivity has declined in the low-clay soils of northern Thailand. Farmers initially responded by adding organic matter from termite mounds, but this was unsustainable in the long-term. Scientists experimented with adding bentonite, one of the smectite family of clays, to the soil. In field trials, conducted by scientists from the International Water Management Institute in cooperation with Khon Kaen University and local farmers, this had the effect of helping retain water and nutrients. Supplementing the farmer's usual practice with a single application of 200 kg bentonite per rai (6.26 rai = 1 hectare) resulted in an average yield increase of 73%. More work showed that applying bentonite to degraded sandy soils reduced the risk of crop failure during drought years.

In 2008, three years after the initial trials, IWMI scientists conducted a survey among 250 farmers in northeast Thailand, half of whom had applied bentonite to their fields. The average improvement for those using the clay addition was 18% higher than for non-clay users. Using the clay had enabled some farmers to switch to growing vegetables, which need more fertile soil. This helped to increase their income. The researchers estimated that 200 farmers in northeast Thailand and 400 in Cambodia had adopted the use of clays, and that a further 20,000 farmers were introduced to the new technique.

If the soil is too high in clay, adding gypsum, washed river sand and organic matter will balance the composition. Adding organic matter (like ramial chipped wood for instance) to soil which is depleted in nutrients and too high in sand will boost its quality.

REFERENCES

1. Voroney, R. Paul; Heck, Richard J. (2015). "The soil habitat". In Paul, Eldor A. Soil microbiology, ecology and biochemistry (4th ed.). Amsterdam, The Netherlands: Elsevier. doi:10.1016/B978-0-12-415955-6.00002-5. ISBN 978-0-12-415955-6.
2. Danoff-Burg, James A. "The terrestrial influence: geology and soils". Earth Institute Center for Environmental Sustainability. New York, New York: Columbia University Press. Retrieved 25 September 2016.
3. Taylor, Sterling A.; Ashcroft, Gaylen L. (1972). Physical edaphology: the physics of irrigated and nonirrigated soils. San Francisco, California: W.H. Freeman. ISBN 978-0716708186.
4. McCarthy, David F. (2006). Essentials of soil mechanics and foundations: basic geotechnics (7th ed.). Upper Saddle River, New Jersey: Prentice Hall. ISBN 978-0131145603.
5. Gilluly, James; Waters, Aaron Clement; Woodford, Alfred Oswald (1975). Principles of geology (4th ed.). San Francisco, California: W.H. Freeman. ISBN 978-0716702696.
6. Ponge, Jean-François (2015). "The soil as an ecosystem" (PDF). Biology and Fertility of Soils. 51 (6): 645–648. doi:10.1007/s00374-015-1016-1.
7. "Soil density". Argonne National Laboratory. Retrieved 25 September 2016.
8. Buol, Stanley W.; Southard, Randal J.; Graham, Robert C.; McDaniel, Paul A. (2011). Soil genesis and classification (6th ed.). Ames, Iowa: Wiley-Blackwell. ISBN 978-0470960608.
9. Retallack, Gregory J. (2008). Soils of the past: an introduction to paleopedology (2nd ed.). Hoboken, New Jersey: John Wiley & Sons. ISBN 978-0470698167.

10. "Glossary of Terms in Soil Science". Agriculture and Agri-Food Canada. Retrieved 9 October 2016.
11. Amundson, Ronald. "Soil preservation and the future of pedology" (PDF). Faculty of Natural Resources. Prince of Songkla University, Songkhla, Thailand. Retrieved 9 October 2016.
12. Simonson, Roy W. (1957). "What soils are". The yearbook of agriculture 1957(PDF) (1st ed.). Washington, D.C.: United States Government Printing Office.
13. Raloff, Janet (17 July 2008). "Dirt is not soil". ScienceNews. Retrieved 23 October 2016.
14. Pouyat, Richard; Groffman, Peter; Yesilonis, Ian; Hernandez, Luis (2002). "Soil carbon pools and fluxes in urban ecosystems" (PDF). Environmental Pollution. 116 (Supplement 1): S107–S118. doi:10.1016/S0269-7491(01)00263-9.
15. Davidson, Eric A.; Janssens, Ivan A. (2006). "Temperature sensitivity of soil carbon decomposition and feedbacks to climate change" (PDF). Nature. 440 (9 March 2006): 165-173. doi:10.1038/nature04514.
16. Powlson, David (2005). "Climatology: will soil amplify climate change?". Nature. 433 (20 January 2005): 204-205. doi:10.1038/433204a. (subscription required (help)).
17. Bradford, Mark A.; Wieder, William R.; Bonan, Gordon B.; Fierer, Noah; Raymond, Peter A.; Crowther, Thomas W. (2016). "Managing uncertainty in soil carbon feedbacks to climate change" (PDF). Nature Climate Change. 6 (27 July 2016): 751–758. doi:10.1038/nclimate3071.
18. Dominati, Estelle; Patterson, Murray; Mackay, Alec (2010). "A framework for classifying and quantifying the natural capital and ecosystem services of soils"(PDF). Ecological Economics. 69 (9): 1858-1868. doi:10.1016/j.ecolecon.2010.05.002.
19. Dykhuizen, Daniel E. (1998). "Santa Rosalia revisited: why are there so many species of bacteria?" (PDF). Antonie van Leeuwenhoek. 73 (1): 25-33. doi:10.1023/A:1000665216662.
20. Torsvik, Vigdis; Øvreås, Lise (2002). Microbial diversity and function in soil: from genes to ecosystems (PDF). Current Opinion in Microbiology. 5 (3): 240-245. doi:10.1016/S1369-5274(02)00324-7.
21. Raynaud, Xavier; Nunan, Naoise (2014). Spatial ecology of bacteria at the microscale in soil (PDF). PLOS ONE. 9 (1): e87217. doi:10.1371/journal.pone.0087217.
22. Whitman, William B.; Coleman, David C.; Wiebe, William J. (1998). Prokaryotes: the unseen majority (PDF). Proceedings of the National Academy of Sciences of the USA. 95 (12): 6578-6583. doi:10.1073/pnas.95.12.6578.
23. Schlesinger, William H.; Andrews, Jeffrey A. (2000). Soil respiration and the global carbon cycle (PDF). Biogeochemistry. 48 (1): 7-20. doi:10.1023/A:1006247623877.
24. Denmead, Owen Thomas; Shaw, Robert Harold (1962). Availability of soil water

to plants as affected by soil moisture content and meteorological conditions(PDF). Agronomy Journal. 54 (5): 385-390. doi:10.2134/agronj1962.00021962005400050005x.

25. House, Christopher H.; Bergmann, Ben A.; Stomp, Anne-Marie; Frederick, Douglas J. (1999). Combining constructed wetlands and aquatic and soil filters for reclamation and reuse of water (PDF). Ecological Engineering. 12 (1–2): 27–38. doi:10.1016/S0925-8574(98)00052-4.
26. Van Bruggen, Ariena H.C.; Semenov, Alexander M. (2000). In search of biological indicators for soil health and disease suppression (PDF). Applied Soil Ecology. 15 (1): 13–24. doi:10.1016/S0929-1393(00)00068-8.
27. A citizen›s guide to monitored natural attenuation (PDF). Retrieved 27 November 2016.
28. Linn, Daniel Myron; Doran, John W. (1984). Effect of water-filled pore space on carbon dioxide and nitrous oxide production in tilled and nontilled soils (PDF). Soil Science Society of America Journal. 48 (6): 1267–1272. doi:10.2136/sssaj1984.03615995004800060013x.
29. Miller, Raymond W.; Donahue, Roy Luther (1990). Soils: an introduction to soils and plant growth. Upper Saddle River, New Jersey: Prentice Hall. ISBN 978-0138202262.
30. Bot, Alexandra; Benites, José (2005). The importance of soil organic matter: key to drought-resistant soil and sustained food and production (PDF). Rome, Italy: Food and Agriculture Organization of the United Nations. ISBN 92-5-105366-9. Retrieved 27 November 2016.
31. Barber, Stanley A. (1995). Chemistry of soil-nutrient associations. In Barber, Stanley A. Soil nutrient bioavailability: a mechanistic approach (2nd ed.). New York, New York: John Wiley & Sons. pp. 9–48. ISBN 978-0-471-58747-7.
32. Möller, Detlev (2015). Chemistry under environmental conditions. In Möller, Detlev. Chemistry for environmental scientists (1st ed.). Berlin, Germany: Walter de Gruyter. pp. 7–52. doi:10.1515/9783110410013-004. ISBN 978-3-11-041001-3.
33. Arizona Master Gardener Manual. Cooperative Extension, College of Agriculture, University of Arizona. Retrieved 4 December 2016.
34. Vannier, Guy (1987). The porosphere as an ecological medium emphasized in Professor Ghilarov›s work on soil animal adaptations. Biology and Fertility of Soils. 3 (1): 39-44. doi:10.1007/BF00260577.
35. Torbert, H. Allen; Wood, Wes (1992). Effect of soil compaction and water-filled pore space on soil microbial activity and N losses (PDF). Communications in Soil Science and Plant Analysis. 23 (11): 1321-1331. doi:10.1080/00103629209368668.
36. Simonson 1957, p. 17.
37. Fookes, Peter G. (1997). Tropical residual soils. London, United Kingdom: Geological Society of London. ISBN 978-1897799383.

38. Soil and water. Food and Agriculture Organization of the United Nations. Retrieved 11 December 2016.
39. Sources. Negative Charge:. Jan.ucc.nau.edu. Retrieved 7 November 2012.
40. Donahue, Miller & Shickluna 1977, p. 4.
41. Kellogg 1957, p. 1.
42. Brady, Nyle (1984). The Nature and Properties of Soils (9th ed.). USA: Macmillan Publishing Co. pp. 4–7. ISBN 0023133406.
43. Kellogg 1957, p. 3.
44. Kellogg 1957, p. 2.
45. Kellogg 1957, p. 4.
46. nitrogen fixation. Encyclopædia Britannica. Encyclopædia Britannica 2009 Deluxe Edition. Chicago: Encyclopædia Britannica, 2009.
47. Kellogg 1957, pp. 1–4.
48. Michael E. Ritter. Factors Affecting Soil Development, Soil Systems, The Physical Environment: an Introduction to Physical Geography, University of Wisconsin, Stevens Point, 1 October 2009, retrieved 3 January 2012.
49. Soils. Natural Resource Conservation Service. United States Department of Agriculture. Retrieved 26 May 2013.
50. Van Schöll, Laura; Smits, Mark M.; Hoffland, Ellis (2006). Ectomycorrhizal weathering of the soil minerals muscovite and hornblende. New Phytologist. 171 (4): 805–814. doi:10.1111/j.1469-8137.2006.01790.x. PMID 16918551.
51. Donahue, Miller & Shickluna 1977, pp. 20–21.
52. Donahue, Miller & Shickluna 1977, p. 21.
53. Donahue, Miller & Shickluna 1977, p. 24.
54. Gilluly, Waters, Woodford (1975). Principles of Geology (4th ed.). USA: W.H. Freeman. p. 215. ISBN 978-0716702696.
55. Donahue, Miller & Shickluna 1977, pp. 28–31.
56. Donahue, Miller & Shickluna 1977, pp. 31–33.
57. H.B. Milford, A.J.E. McGaw and K.J. Nixon, Soil Data Entry Handbook for the NSW Soil and Land Information System (SALIS), 3rd ed., New South Wales Department of Land and Water Conservation Resource Information Systems Group, Parramatta, 2001, pdf pp. 30–32.
58. Gove Hambidge, Climate and Man—A Summary, in Erwin Raisz, U.S. Department of Agriculture, Climate And Man, Part One, Yearbook of Agriculture 1941, Repr. Honolulu: University Press of the Pacific, 2004, ISBN 978-1-4102-1538-3, pp. 1–66, p. 27.
59. Donahue, Miller & Shickluna 1977, p. 35.
60. Copley, Jon (25 August 2005). Millions of bacterial species revealed underfoot. New Scientist. Retrieved 19 April 2010.
61. Amber Dance (2008). Soil ecology: What lies beneath. Nature. 455 (7214):

724–5. doi:10.1038/455724a. PMID 18843336.

62. Roesch LF, Fulthorpe RR, Riva A, Casella G, Hadwin AK, Kent AD, Daroub SH, Camargo FA, Farmerie WG, Triplett EW; Fulthorpe; Riva; Casella; Hadwin; Kent; Daroub; Camargo; Farmerie; Triplett (2007). Pyrosequencing enumerates and contrasts soil microbial diversity. The ISME Journal. 1 (4): 283–90. doi:10.1038/ismej.2007.53. PMC 2970868 . PMID 18043639.
63. Gans J, Wolinsky M, Dunbar J.; Wolinsky; Dunbar (2005). Computational improvements reveal great bacterial diversity and high metal toxicity in soil. Science. 309 (5739): 1387–90. Bibcode:2005Sci...309.1387G. doi:10.1126/science.1112665. PMID 16123304.
64. Simonson 1957, pp. 20, 21.
65. Donahue, Miller & Shickluna 1977, p. 26.
66. Nyle C. Brady & Ray R. Weil (2009). Elements of the Nature and Properties of Soils (3rd Edition). Prentice Hall. ISBN 9780135014332.
67. R. B. Brown (September 2003). Soil Texture (PDF). Fact Sheet SL-29. University of Florida, Institute of Food and Agricultural Sciences. Retrieved 8 July 2008.
68. Russell 1957, pp. 32–33.
69. Flemming 1957, p. 331.
70. Donahue, Miller & Shickluna 1977, p. 53.
71. Soil Survey Division Staff (1993). Soil Structure. Handbook 18. Soil survey manual. Soil Conservation Service. U.S. Department of Agriculture. Archived from the original on 16 March 2008. Retrieved 8 July 2008.
72. Donahue, Miller & Shickluna 1977, pp. 55–56.
73. Soil Structure - Physical Properties. The Cooperative Soil Survey. Retrieved 19 October 2012.
74. Donahue, Miller & Shickluna 1977, pp. 59–61.
75. Donahue, Miller & Shickluna 1977, p. 60.
76. Physical aspects of crop productivity. www.fao.org. FAO. Retrieved 2 July 2016.
77. Donahue, Miller & Shickluna 1977, pp. 62–63.
78. Soils - Part 2: Physical Properties of Soil and Soil Water. Retrieved 19 October 2012.
79. Johnson, T.A., T.R. Ellsworth, R.J.M. Hudson, and G.K. Sims. 2013. Diffusion Limitation for Atrazine Biodegradation in Soil. Advances in Microbiology. 3(5): 412-420.
80. Donahue, Miller & Shickluna 1977, pp. 62–63,565–567.
81. Taylor, H.M. 1983. Managing root systems for efficient water use: an overview, p. 87–113 in Taylor, H.H. et al. (Eds.). Limitations to Efficient Water Use in Crop Production. ASA, CCSA, and SSSA, Madison WI.
82. Gill, D. 1975. Influence of white spruce trees on permafrost-table microtopography, Mackenzie River Delta. Can. J. Earth Sci. 12(2):263–272.

83. Mityga, H.G.; Lanphear, F.O. 1971. Factors influencing the cold hardiness of Taxus cuspidata roots. J. Amer. Soc. Hort. Sci. 96:83–86.

84. Havis, J.R. 1976. Root hardiness of woody ornamentals. HortScience 11:385–386.

85. Lyr, H.; Hoffmann, G. 1967. Growth rates and growth periodicity of tree roots. p. 181–236 in Romberger, J.A.; Mikola, P. (Eds.). International Review of Forest Research, Vol. 2, Academic Press, New York NY.

86. Chalupa, V.; Fraser, D.A. 1968. Effect of soil and air temperature on soluble sugars and growth of white spruce (Picea glauca) seedlings. Can. J. Bot. 46:65–69.

87. Heninger, R.L.; White, D.P. 1974. Tree seedling growth at different soil temperatures. For. Sci. 20:363–367.

88. Ritchie, G.A.; Dunlap, J.R. 1980. Root growth potential: its development and expression in forest tree seedlings. New Zealand J. For. Sci. 10:218–248.

89. Binder, W.D.; Spittlehouse, D.L.; Draper, D.A. 1988. Post-planting studies of the physiology of white spruce 1984–1985, E.P. 966. B.C. Min. For., Res. Branch, Victoria BC, Progr. Rep. 5, unpubl. 85 p. [Coates et al. 1994 give date as 1987, Binder bibliog at R986 gives 1988]

90. Landhäusser, S.M.; DesRochers, A.; and Lieffers, V.J. 2001. A comparison of growth and physiology in Picea glauca and Populus tremuloides at different soil temperatures. Can. J. For. Res. 31:1922–1929.

91. Tryon, P.R.; Chapin, F.S. 1983. Temperature control over root growth and root biomass in taiga forest trees. Can. J. For. Res. 13(5):827–833.

92. Landhäusser, S.M.; Silins, U.; Lieffers, V.J.; Liu, W. 2003. Response of Populus tremuloides, Populus balsamifera, Betula papyrifera and Picea glauca seedlings to low soil temperature and water-logged soil conditions. Scan. J. For. Res. 18:391–400. (Citcd Green 2004)

93. Turner, N.C.; Jarvis, P.G. 1975. Photosynthesis in Sitka spruce (Picea sitchensis Bong.). IV. Response to soil temperature. J. Appl. Ecol. 12:561–576.

94. Day, T.A.; DeLucia, E.H.; Smith, W.K. 1990. Effect of soil temperature on stem flow, shoot gas exchange and water potential of Picea engelmannii (Parry) during snowmelt. Oecologia 84(4):474–481.

95. Green, D.S. 2004. Describing condition-specific determinants of competition in boreal and sub-boreal mixedwood stands. For. Chron. 80(6):736–742.

96. Donahue, Miller & Shickluna 1977, p. 71.

97. Arizona Master Gardener Manual. Cooperative Extension, College of Agriculture, University of Arizona. p. Chapter 2, pp 4–8. Retrieved 27 May 2013.

98. The Color of Soil. United States Department of Agriculture - Natural Resources Conservation Service. Archived from the original on 16 March 2008. Retrieved 8 July 2008.

99. Electrical Design, Cathodic Protection. United States Army Corps of Engineers. 22 April 1985. Archived from the original on 12 June 2008. Retrieved 2 July 2008.

100. The why and how to testing the Electrical Conductivity of Soils | Resources. Retrieved 19 December 2010.
101. R. J. Edwards (15 February 1998). Typical Soil Characteristics of Various Terrains. Retrieved 2 July 2008.
102. Donahue, Miller & Shickluna 1977, p. 72.
103. Wadleigh 1957, p. 48.
104. Richards & Richards 1957, p. 50.
105. Richards & Richards 1957, p. 56.
106. Wadleigh 1957, p. 39.
107. Richards & Richards 1957, p. 52.
108. Donahue, Miller & Shickluna 1977, pp. 72-74.
109. CHAPTER 2 - SOIL AND WATER. Fao.org. Retrieved 7 November 2012.
110. Donahue, Miller & Shickluna 1977, pp. 72–75.
111. Wadleigh 1957, p. 47.
112. Donahue, Miller & Shickluna 1977, pp. 75–76.
113. Donahue, Miller & Shickluna 1977, pp. 76–77.
114. Donahue, Miller & Shickluna 1977, p. 80.
115. Donahue, Miller & Shickluna 1977, p. 85.
116. Donahue, Miller & Shickluna 1977, p. 86.
117. Donahue, Miller & Shickluna 1977, p. 88.
118. Brehm, Denise (11 December 2008). CEE researchers explain mystery of gravity fingers. MIT Department of Civil & Environmental Engineering. MIT. Retrieved 31 October 2012.
119. Urban Trees Enhance Water Infiltration. Fisher, Madeline. The American Society of Agronomy. 17 November 2008. Retrieved 31 October 2012.
120. Major floods recharge aquifers. University of New South Wales Science. 24 January 2011. Retrieved 31 October 2012.
121. Donahue, Miller & Shickluna 1977, p. 90.
122. Donahue, Miller & Shickluna 1977, p. 91.
123. Donahue, Miller & Shickluna 1977, p. 92.
124. Wadleigh 1957, p. 46.
125. Donahue, Miller & Shickluna 1977, p. 94.
126. Donahue, Miller & Shickluna 1977, pp. 97–99.
127. Russell 1957, pp. 35–36.
128. http://atmo.tamu.edu/class/geos489/lecture6/soiltracegasrev.pdf
129. College of Tropical Agriculture and Human Resources. Soil Mineralogy. www.ctahr.hawaii.edu. University of Hawai'i. Retrieved 7 August 2014.
130. Russell, E. Walter (1973). Soil conditions and plant growth. (10th ed.). London:

Longman. pp. 67–70. ISBN 0582440483.
131. Donahue, Miller & Shickluna 1977, pp. 101–102.
132. Simonson 1957, p. 19.
133. Donahue, Miller & Shickluna 1977, p. 102.
134. Russell 1957, p. 33.
135. Donahue, Miller & Shickluna 1977, pp. 102–107.
136. Donahue, Miller & Shickluna 1977, pp. 101–107.
137. Donahue, Miller & Shickluna 1977, p. 107.
138. Donahue, Miller & Shickluna 1977, p. 108.
139. Russell 1957, pp. 33, 34.
140. Coleman & Mehlich 1957, p. 74.
141. Donahue, Miller & Shickluna 1977, pp. 108–110.
142. Dean 1957, p. 82.
143. Allison 1957, p. 90.
144. Reitemeier 1957, p. 103.
145. Donahue, Miller & Shickluna 1977, p. 110.
146. Coleman & Mehlich 1957, p. 73.
147. Holmes & Brown 1957, p. 112.
148. Donahue, Miller & Shickluna 1977, p. 111.
149. Olsen & Fried 1957, p. 96.
150. Reitemeier 1957, p. 101.
151. Donahue, Miller & Shickluna 1977, pp. 103–112.
152. Simonson 1957, pp. 18, 21–24, 29.
153. Russell 1957, pp. 32, 35.
154. Donahue, Miller & Shickluna 1977, p. 112.
155. Russell 1957, p. 35.
156. Allaway 1957, p. 69.
157. Lehmann, J. Terra Preta de Indio. University of Cornell, Dept. of Crop and Soil Sciences. Retrieved 30 March 2013.
158. Donahue, Miller & Shickluna 1977, p. 103–106.
159. Donahue, Miller & Shickluna 1977, p. 114.
160. Donahue, Miller & Shickluna 1977, pp. 115–116.
161. Chang, Raymond (1984). Chemistry. Random House, Inc. p. 424. ISBN 0-394-32983-X.
162. Donahue, Miller & Shickluna 1977, pp. 116–117.
163. Donahue, Miller & Shickluna 1977, pp. 116–119.
164. Donahue, Miller & Shickluna 1977, pp. 119–120.

165. Donahue, Miller & Shickluna 1977, pp. 120–121.

166. Dean 1957, p. 80.

167. Russel 1957, pp. 123–125.

168. Dean 1957, pp. 80, 81.

169. Roy, R.N.; Finck, A.; Blair, G.J.; Tandon, H.L.S. (2006). Chapter 4: Soil fertility and crop production. Plant nutrition for food security: a guide for integrated nutrient management (PDF). Rome: Food and Agriculture Organization of the United Nations. pp. 43–90. ISBN 92-5-105490-8. Retrieved 20 June 2016.

170. Donahue, Miller & Shickluna 1977, pp. 123–131.

171. Donahue, Miller & Shickluna 1977, p. 125.

172. Donahue, Miller & Shickluna 1977, p. 126.

173. Lecture 22. Northern Arizona University. Retrieved 22 March 2013.

174. Donahue, Miller & Shickluna 1977, pp. 123–128.

175. Wadleigh 1957, p. 41.

176. Broadbent 1957, p. 153.

177. Donahue, Miller & Shickluna 1977, p. 128.

178. Allison 1957, pp. 85–94.

179. Broadbent 1957, pp. 152–155.

180. Donahue, Miller & Shickluna 1977, pp. 128–131.

181. Donahue, Miller & Shickluna 1977, pp. 129–130.

182. Donahue, Miller & Shickluna 1977, p. 145.

183. Donahue, Miller & Shickluna 1977, pp. 128–129.

184. Allison 1957, p. 87.

185. Donahue, Miller & Shickluna 1977, p. 130.

186. Donahue, Miller & Shickluna 1977, p. 131.

187. Donahue, Miller & Shickluna 1977, pp. 134–135.

188. Reitemeier 1957, pp. 101–104.

189. Donahue, Miller & Shickluna 1977, pp. 135–136.

190. Donahue, Miller & Shickluna 1977, p. 136.

191. Jordan & Reisenauer 1957, pp. 107.

192. Holmes & Brown 1957, pp. 111.

193. Sherman 1957, pp. 135.

194. Seatz & Jurinak 1957, pp. 115.

195. Reuther 1957, pp. 128.

196. Russel 1957, pp. 121.

197. Stout & Johnson 1957, pp. 146.

198. Stout & Johnson 1957, pp. 141.

199. Donahue, Miller & Shickluna 1977, pp. 136–137.

200. Stout & Johnson 1957, pp. 107.

201. Pimentel, D.; et al. (1995). Environmental and economic costs of soil erosion and conservation benefits. Science. 267 (24): 1117–22. Bibcode:1995Sci...267.1117P. doi:10.1126/science.267.5201.1117. PMID 17789193.

202. Foth, Henry D. (1984). Fundamentals of soil science. New York: Wiley. p. 151. ISBN 0-471-88926-1.

203. Gilluly, Waters, Woodford (1975). Principles of Geology (4th ed.). USA: W.H. Freeman. p. 216. ISBN 978-0716702696.

204. Verkaik, Eric; Jongkind, Anne G.; Berendse, Frank (2006). Short-term and long-term effects of tannins on nitrogen mineralization and litter decomposition in kauri (Agathis australis (D. Don) Lindl.) forests. Plant and Soil. 287: 337. doi:10.1007/s11104-006-9081-8.

205. Fierer, N.; Schimel, Joshua P.; Cates, Rex G.; Zou, Jiping (2001). Influence of balsam poplar tannin fractions on carbon and nitrogen dynamics in Alaskan taiga floodplain soils. Soil Biology and Biochemistry. 33 (12–13): 1827. doi:10.1016/S0038-0717(01)00111-0.

206. Wagai, Rota; Mayer, Lawrence M.; Kitayama, Kanehiro; Knicker, Heike (2008). Climate and parent material controls on organic matter storage in surface soils: A three-pool, density-separation approach. Geoderma. 147: 23–33. doi:10.1016/j.geoderma.2008.07.010.

207. Minayeva, T. Yu.; Trofimov, S. Ya.; Chichagova, O.A.; Dorofeyeva, E.I.; Sirin, A.A.; Glushkov, I.V.; Mikhailov, N.D.; Kromer, B. (2008). Carbon accumulation in soils of forest and bog ecosystems of southern Valdai in the Holocene. Biology Bulletin. 35 (5): 524–532. doi:10.1134/S1062359008050142.

208. Sanchez, Pedro A. (1976). Properties and management of soils in the tropics. New York: Wiley. ISBN 0471752002.

209. Paul, E.A. (1997). Soil organic matter in temperate agroecosystems : long-term experiments in North America. Boca Raton: CRC Press. p. 80. ISBN 9780849328022.

210. Retallack, G. J. (1990). Soils of the past : an introduction to paleopedology. Boston: Unwin Hyman. p. 32. ISBN 978-0-04-445757-2.

211. Buol, S.W. (1990). Soil genesis and classification. Ames, Iowa: Iowa State University Press. p. 36. doi:10.1081/E-ESS. ISBN 0-8138-2873-2.

212. IUSS Working Group WRB (2014). World Reference Base for Soil Resources 2014. International soil classification system for naming soils and creating legends for soil maps (PDF) (3rd ed.). Rome: FAO. ISBN 978-92-5-108370-3. Retrieved 29 August 2014.

213. bhttp://eusoils.jrc.ec.europa.eu/esdb_archive/eusoils_docs/other/EUR23439.pdfSoils of the European Union by the EU Institute for Environment and Sustainability. Accessed on 8 October 2013

214. The Soil Orders, Department of Environmental Sciences, University of Virginia,

retrieved 23 October 2012.

215. Donahue, Miller & Shickluna 1977, pp. 411–432.

216. Soil Survey Staff (1999). Soil taxonomy: A Basic System of Soil Classification for Making and Interpreting Soil Surveys. 2nd edition. Natural Resources Conservation Service. U.S. Department of Agriculture Handbook 436. (PDF). USDA: United States Dept. of Agriculture, Naturel Resources Conservation Service. Retrieved 10 March 2013.

217. The Twelve Soil Orders: Soil Taxonomy, Soil & Land Resources Division, College of Agricultural and Life Sciences, University of Idaho

218. Donahue, Miller & Shickluna 1977, p. 409.

219. Donahue, Miller & Shickluna 1977, pp. 409-410.

220. Donahue, Miller & Shickluna 1977, p. 410.

221. Leake, Simon; Haege, Elke (2014). Soils for Landscape Development. CSIRO Publishing. ISBN 9780643109643.

222. Ponge, Jean-François (2003). Humus forms in terrestrial ecosystems: a framework to biodiversity (PDF). Soil Biology and Biochemistry. 35 (7): 935–945. doi:10.1016/S0038-0717(03)00149-4.

223. De Deyn, Gerlinde B.; Van der Putten, Wim H. (2005). Linking aboveground and belowground diversity. Trends in Ecology & Evolution. 20 (11): 625–633. doi:10.1016/j.tree.2005.08.009. PMID 16701446.

224. Hansen, J.; et al. (2008). Target atmospheric CO2: Where should humanity aim?. Open Atmospheric Science Journal. 2: 217–31. arXiv:0804.1126 . Bibcode:2008OASJ....2..217H. doi:10.2174/1874282300802010217

225. Lal, R. (11 June 2004). Soil Carbon Sequestration Impacts on Global Climate Change and Food Security. Science. 304 (5677): 1623–1627. Bibcode:2004Sci...304.1623L. doi:10.1126/science.1097396. PMID 15192216.

226. Blakeslee, Thomas R. (24 February 2010). Greening Deserts for Carbon Credits. Renewable Energy World.com. Retrieved 23 October 2012.

227. Setz, EZF; Enzweiler J; Solferini VN; Amendola MP; Berton RS (1999). Geophagy in the golden-faced saki monkey (Pithecia pithecia chrysocephala) in the Central Amazon. Journal of Zoology. 247 (1): 91–103. doi:10.1111/j.1469-7998.1999.tb00196.x. (subscription required (help)).

228. Kohne, John Maximilian; Koehne, Sigrid; Simunek, Jirka (2009). A review of model applications for structured soils: a) Water flow and tracer transport. Journal of Contaminant Hydrology. 104 (1–4): 4–35. Bibcode:2009JCHyd.104....4K. doi:10.1016/j.jconhyd.2008.10.002. PMID 19012994.

229. Diplock, EE; Mardlin DP; Killham KS; Paton GI (2009). Predicting bioremediation of hydrocarbons: laboratory to field scale. Environmental Pollution. 157 (6): 1831–1840. doi:10.1016/j.envpol.2009.01.022. PMID 19232804.

230. Moeckel, Claudia; Nizzetto, Luca; Di Guardo, Antonio; Steinnes, Eiliv; Freppaz, Michele; Filippa, Gianluca; Camporini, Paolo; Benner, Jessica; Jones, Kevin C. (2008). Persistent organic pollutants in boreal and montane soil profiles: distribution,

evidence of processes and implications for global cycling. Environmental Science and Technology. 42 (22): 8374–8380. Bibcode:2008EnST...42.8374M. doi:10.1021/es801703k. PMID 19068820.

231. Rezaei, Khalil; Guest, Bernard; Friedrich, Anke; Fayazi, Farajollah; Nakhaei, Mohamad; Aghda, Seyed Mahmoud Fatemi; Beitollahi, Ali (2009). Soil and sediment quality and composition as factors in the distribution of damage at the December 26, 2003, Bam area earthquake in SE Iran (M (s)=6.6). Journal of Soils and Sediments. 9: 23–32. doi:10.1007/s11368-008-0046-9.

232. Johnson, D.L.; Ambrose, S.H.; Bassett, T.J.; Bowen, M.L.; Crummey, D.E.; Isaacson, J.S.; Johnson, D.N.; Lamb, P.; Saul, M.; Winter-Nelson, A. E. (1997). Meanings of environmental terms. Journal of Environmental Quality. 26 (3): 581–589. doi:10.2134/jeq1997.00472425002600030002x.

233. Jones, j. a. a. (1976). Soil piping and stream channel initiation. Water Resources Research. 7 (3): 602–610. Bibcode:1971WRR.....7..602J. doi:10.1029/WR007i003p00602.

234. Dooley, Alan (June 2006). Sandboils 101: Corps has experience dealing with common flood danger. Engineer Update. US Army Corps of Engineers. Archived from the original on 18 April 2008. Retrieved 14 May 2008.

235. ILRI (1989). Effectiveness and Social/Environmental Impacts of Irrigation Projects: a Review (pdf). In: Annual Report 1988 of the International Institute for Land Reclamation and Improvement (ILRI). Wageningen, The Netherlands. pp. 18–34.

236. Drainage Manual: A Guide to Integrating Plant, Soil, and Water Relationships for Drainage of Irrigated Lands. Interior Dept., Bureau of Reclamation. 1993. ISBN 0-16-061623-9.

237. Free articles and software on drainage of waterlogged land and soil salinity control. Retrieved 28 July 2010.

238. Improving soils and boosting yields in Thailand (PDF). Success Stories. International Water Management Institute (2). 2010. doi:10.5337/2011.0031.

239. Provide for your garden›s basic needs ... and the plants will take it from there. USA Weekend. 10 March 2011.

Soil Chemistry

Chapter Contents

- **Soil Chemistry**
- **Concepts**
- **Soil pH**
- **Soil Carbon**
- **Soil Gas**

Soils are a complex mixture of organic and inorganic, living and non-living components. And with this complex mix comes some very complex – yet important – chemistry. If there is one principle that can frame and help simplify soil chemistry, it is pH. The influence of pH over living and non-living processes has earned pH the 'master variable' in soil science: change pH, and you likely alter or influence a slew of processes. The reason for this is that pH is intimately tied to chemical, biochemical, and biological processes, and their interactions.

SOIL CHEMISTRY

Soil chemistry is the study of the chemical characteristics of soil. Soil chemistry is affected by mineral composition, organic matter and environmental factors.

Soil chemistry is the study of the chemical characteristics of soil. Soil chemistry is affected by mineral composition, organic matter and environmental factors. Soil - (i) The unconsolidated mineral or organic material on the immediate surface of the Earth that serves as a natural medium for the growth of land plants. (ii) The unconsolidated mineral or organic matter on the surface of the Earth that has been subjected to and shows effects of genetic and environmental factors of: climate (including water and temperature effects), and macro- and microorganisms, conditioned by relief, acting on parent material over a period of time. A product-soil differs from the material from which it is derived in many physical, chemical,biological, and morphological properties and characteristics. This definition is from Soil Taxonomy, second edition.

Key facts you should know about Soil:

Soil makes up the outermost layer of our planet and is formed from rocks and decaying plants and Animals. Soil has varying amounts of organic matter (living and dead organisms), minerals, and nutrients.

- An average soil sample is 45 percent minerals, 25 percent water, 25 percent air, and five percent organic matter. Different-sized mineral particles, such as sand, silt, and clay, give soil its texture.
- Topsoil is the most productive soil layer.
- Ten tonnes of topsoil spread evenly over a hectare is only as thick as a one Euro coin.
- Natural processes can take more than 500 years to form 2 centimeters of topsoil.
- In some cases, 5 tonnes of animal life can live in one hectare of soil.
- Fungi and bacteria help break down organic matter in the soil.
- Earthworms digest organic matter, recycle nutrients, and make the surface soil richer.
- Roots loosen the soil, allowing oxygen to penetrate. This benefits animals living in the soil. They also benefit roots which require oxygen themselves. They further hold soil together and help prevent erosion.
- A fully functioning soil reduces the risk of floods and protects underground water supplies by neutralizing or filtering out potential pollutants and storing as much as 3750 tonnes of water per hectare.
- Scientists have found 10,000 types of soil in Europe and about 70,000 types of soil in the United States.
- Soil stores 10% of the world›s carbon dioxide emissions.

Environmental Soil Chemistry

A knowledge of environmental soil chemistry is paramount to predicting the fate of contaminants, as well as the processes by which they are initially released into the soil. Once a chemical is exposed to the soil environment myriad chemical reactions can occur that may increase or decrease contaminant toxicity. These reactions include adsorption/desorption, precipitation, polymerization, dissolution, complexation and oxidation/reduction. These reactions are often disregarded by scientists and engineers involved with environmental remediation. Understanding these processes enable us to better predict the fate and toxicity of contaminants and provide the knowledge to develop scientifically correct, and cost-effective remediation strategies.

CONCEPTS

Anion and cation exchange capacity

Soil pH

Mineral formation and transformation processes

Clay mineralogy

Sorption and precipitation reactions in soil

Oxidation-reduction reactions

Chemistry of problem soils

SOIL pH

The **soil pH** is a measure of the acidity or alkalinity in soils. pH is defined as the negative logarithm (base 10) of the activity of hydronium ions (H^+ or, more precisely, $H_3O^+_{aq}$) in a solution. In water, it normally ranges from -1 to 14, with 7 being neutral. A pH below 7 is acidic and above 7 is alkaline. Soil pH is considered a master variable in soils as it controls many chemical processes that take place. It specifically affects plant nutrient availability by controlling the chemical forms of the nutrient. The optimum pH range for most plants is between 5.5 and 7.0, however many plants have adapted to thrive at pH values outside this range.

Classification of Soil Ph Ranges

The United States Department of Agriculture Natural Resources Conservation Service, formerly Soil Conservation Service classifies soil pH ranges as follows:

Denomination	pH range
Ultra acidic	< 3.5
Extremely acidic	3.5–4.4
Very strongly acidic	4.5–5.0
Strongly acidic	5.1–5.5
Moderately acidic	5.6–6.0
Slightly acidic	6.1–6.5
Neutral	6.6–7.3
Slightly alkaline	7.4–7.8
Moderately alkaline	7.9–8.4
Strongly alkaline	8.5–9.0
Very strongly alkaline	> 9.0

Sources of Soil Ph

Sources of Acidity

Acidity in soils comes from H^+ and Al^{3+} ions in the soil solution and sorbed to soil surfaces. While pH is the measure of H^+ in solution, Al^{3+} is important in acid soils because between pH 4 and 6, Al^{3+} reacts with water (H_2O) forming $AlOH^{2+}$, and $Al(OH)_2^+$, releasing extra H^+ ions. Every Al^{3+} ion can create 3 H^+ions. Many other processes contribute to the formation of acid soils including rainfall, fertilizer use, plant root activity and the weathering of primary and secondary soil minerals. Acid soils can also be caused by pollutants such as acid rain and mine spoilings.

- Rainfall: Acid soils are most often found in areas of high rainfall. Excess rainfall leaches base cation from the soil, increasing the percentage of Al^{3+} and H^+relative to other cations. Additionally, rainwater has a slightly acidic pH of 5.7 due to a reaction with CO_2 in the atmosphere that forms carbonic acid.
- Fertilizer use: Ammonium (NH_4^+) fertilizers react in the soil in a process called nitrification to form nitrate (NO_3^-), and in the process release H^+ ions.
- Plant root activity: Plants take up nutrients in the form of ions (NO_3^-, NH_4^+, Ca^{2+}, $H_2PO_4^-$, etc.), and often, they take up more cations than anions. However plants must maintain a neutral charge in their roots. In order to compensate for the extra positive charge, they will release H^+ ions from the root. Some plants will also exude organic acids into the soil to acidify the zone around their roots to help solubilize metal nutrients that are insoluble at neutral pH, such as iron (Fe).
- Weathering of minerals: Both primary and secondary minerals that compose soil contain Al. As these minerals weather, some components such as Mg, Ca, and K, are taken up by plants, others such as Si are leached from the soil, but due to chemical properties, Fe and Al remain in the soil profile. Highly weathered soils are often characterized by having high concentrations of Fe and Al oxides.
- Acid rain: When atmospheric water reacts with sulfur and nitrogen compounds that result from industrial processes, the result can be the formation of sulfuric and nitric acid in rainwater. However the amount of acidity that is deposited in rainwater is much less, on average, than that created through agricultural activities.
- Mine spoil: Severely acidic conditions can form in soils near mine spoils due to the oxidation of pyrite.
- Potential acid sulfate soils naturally formed in waterlogged coastal and estuarine environments can become highly acidic when drained or excavated.
- Decomposition of organic matter by microorganisms releases CO_2 which when mixed with soil water can form carbonic acid (H_2CO_3).

Sources of Alkalinity

Alkaline soils have a high saturation of base cations (K^+, Ca^{2+}, Mg^{2+} and Na^+). This is due to an accumulation of soluble salts which are classified as either saline soil, sodic soil, saline-sodic soil or alkaline soil. All saline and sodic soils have high salt concentrations, with saline soils being dominated by calcium and magnesium salts and sodic soils being dominated by sodium. Alkaline soils are characterized by the presence of carbonates. Soil in areas with limestone near the surface are alkaline from the calcium carbonate in limestone constantly mixing with the soil. Groundwater sources in these areas contain dissolved limestone.

Effect of Soil Ph on Plant Growth

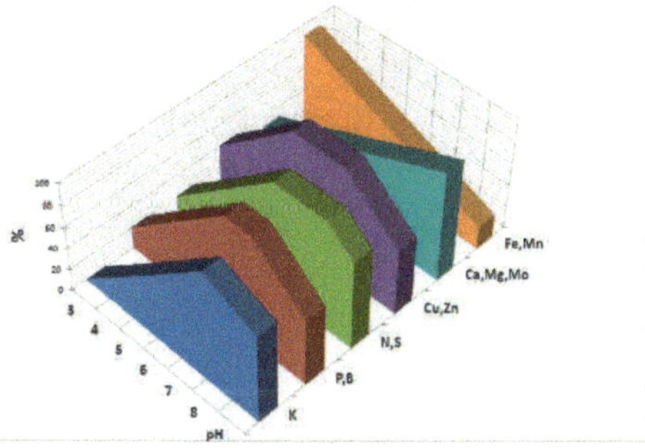

Nutrient availability in relation to soil pH

Acid Affected Soils

Plants grown in acid soils can experience a variety of symptoms including aluminium (Al), hydrogen (H), and/or manganese (Mn) toxicity, as well as nutrient deficiencies of calcium (Ca) and magnesium (Mg).

Aluminium toxicity is the most widespread problem in acid soils. Aluminium is present in all soils, but dissolved Al^{3+} is toxic to plants; Al^{3+} is most soluble at low pH, above pH 5.2 little Al is in soluble form in most soils. Aluminium is not a plant nutrient, and as such, is not actively taken up by the plants, but enters plant roots passively through osmosis. Aluminium inhibits root growth; lateral roots and root tips become thickened and roots lack fine branching; root tips may turn brown. In the root, Al has been shown to interfere with many physiological processes including the uptake and transport of calcium and other essential nutrients, cell division, cell wall formation, and enzyme activity.

Below pH 4, H^+ ions themselves damage root cell membranes.

In soils with high content of manganese-containing minerals, Mn toxicity can become a problem at pH 5.6 and lower. Manganese, like aluminium, becomes increasingly soluble as pH drops, and Mn toxicity symptoms can be seen at pH levels below 5.6. Manganese is an essential plant nutrient, so plants transport Mn into leaves. Classic symptoms of Mn toxicity are crinkling or cupping of leaves.

Nutrient Availability in Relation to Soil Ph

Nutrients needed in large amounts by plants are referred to as macronutrients and include nitrogen (N), phosphorus (P), potassium (K), calcium (Ca), magnesium (Mg) and sulfur (S). Elements that plants need in trace amounts are called trace nutrients or micronutrients. Trace nutrients are not major components of plant tissue but are essential for growth. They include iron (Fe), manganese (Mn), zinc (Zn), copper (Cu), cobalt (Co), molybdenum (Mo), and boron (B). Both macronutrient and micronutrient availability are affected by soil pH. In slightly to moderately alkaline soils, molybdenum and macronutrient (except for phosphorus) availability is increased, but P, Fe, Mn, Zn Cu, and Co levels are reduced and may adversely affect plant growth. In acidic soils, micronutrient availability (except for Mo and Bo) is increased. Nitrogen is supplied as ammonium (NH_4) or nitrate (NO_3) by nitrogen fixation or fertilizer amendments, and dissolved N will have the highest concentrations in soil with pH 6–8. Concentrations of available N are less sensitive to pH than concentration of available P. In order for P to be available for plants, soil pH needs to be in the range 6.0 and 7.5. If pH is lower than 6, P starts forming insoluble compounds with iron (Fe) and aluminium (Al) and if pH is higher than 7.5 P starts forming insoluble compounds with calcium (Ca). Most nutrient deficiencies can be avoided between a pH range of 5.5 to 6.5, provided that soil minerals and organic matter contain the essential nutrients to begin with.

Water Availability in Relation to Soil Ph

Further information: Water content and Water potential

Determining Ph

Methods of determining pH include:

- Observation of soil profile: Certain profile characteristics can be indicators of either acid, saline, or sodic conditions. Strongly acidic soils often have poor incorporation of the organic surface layer with the underlying mineral layer. The mineral horizons are distinctively layered in many cases, with a pale eluvial (E) horizon beneath the organic surface; this E is underlain by a darker B horizon in a classic podzol horizon sequence. This is a very rough gauge of acidity as there is no correlation between thickness of the E and soil pH. E horizons a few feet thick in Florida usually have pH just above 5 (merely strongly acid) while E horizons a few inches thick in New England are extremely acid with pH readings of 4.5 or below. In the southern Blue Ridge Mountains there are ultra acid soils, pH below 3.5, which have no E horizon. Presence of a caliche layer indicates the presence of calcium carbonates, which are present in alkaline conditions. Also, columnar structure can be an indicator of sodic condition.

- Observation of predominant flora. Calcifuge plants (those that prefer an acidic soil) include *Erica*, *Rhododendron* and nearly all other Ericaceae species, many birch (*Betula*), foxglove (*Digitalis*), gorse (*Ulex* spp.), and Scots Pine (*Pinus sylvestris*). Calcicole (lime loving) plants include ash trees (*Fraxinus* spp.), honeysuckle (*Lonicera*), *Buddleja*, dogwoods (*Cornus* spp.), lilac (*Syringa*) and *Clematis* species.
- Use of an inexpensive pH testing kit, where in a small sample of soil is mixed with indicator solution which changes colour according to the acidity/alkalinity.
- Use of litmus paper. A small sample of soil is mixed with distilled water, into which a strip of litmus paper is inserted. If the soil is acidic the paper turns red, if alkaline, blue.
- Use of a commercially available electronic pH meter, in which a rod is inserted into moistened soil and measures the concentration of hydrogen ions.

Examples of Plant Ph Preferences

- pH 4.5–5.0: Ericaceae (azalea, bilberry, blueberry, cranberry; heather), hydrangea for blue, (less acidic for pink), liquidambar or sweet gum, orchid, pin oak
- pH 5.0–5.5: Boronia, daphne; Ericaceae: (camellia, heather, rhododendron), ferns, iris, orchids, parsley, conifers (e.g., pine); Poaceae: (maize, millet, rye, oat), radish; Solanales: (potato, sweet potato); Bromeliaceae (pineapple).
- pH 5.5–6.0: Asteraceae: (aster, endive); Brassicaceae: (brussels sprout, kohlrabi), carrot; Cucurbitales: (begonia, chayote or choko); Fabaceae: (bean, crimson clover, peanut, soybean), petunia, rhubarb, violet, most bulbs (canna, daffodil, jonquil), larkspur, primrose
- pH 6.0–6.5 antirrhinum or snapdragon, Brassicaceae: (broccoli, cabbage, candytuft, cauliflower, turnip, wallflower); Cucurbitaceae: (cucumber, pumpkin, squash); Fabaceae: (pea, red clover, white clover), gladiolus, Iceland poppy; Rosales: (cannabis, rose, strawberry); Solanaceae: (eggplant or aubergine, tomato), sweet corn; Violaceae: (pansy, viola), zinnia or zinnea
- pH 6.5–7.0: Amaranthaceae: (beet, spinach); Apiaceae: (celery, parsnip); Asparagales: (asparagus, onion); Asteraceae: (chrysanthemum, dahlia, lettuce), carnation; Fabaceae: (alfalfa, sweet pea), melon, stock, tulip
- pH 7.1–8.0 lilac

Changing Soil Ph

Increasing Ph of Acidic Soil

The most common amendment to increase soil pH is lime ($CaCO_3$ or $MgCO_3$), usually in the form of finely ground agricultural lime. The amount of lime needed to change pH is determined by the mesh size of the lime (how finely it is ground)and the buffering capacity of the soil. A high mesh size (60–100) indicates a finely ground lime, that will react quickly with soil acidity. Buffering capacity of soils is a function of a soils cation exchange capacity, which is in turn determined by the clay content of the soil, the type of clay and the amount of organic matter present. Soils with high clay content, particularly shrink–swell clay, will have a higher buffering capacity than soils with little clay. Soils with high organic matter will also have a higher buffering capacity than those with low organic matter. Soils with high buffering capacity require a greater amount of lime to be added than a soil with a lower buffering capacity for the same incremental change in pH. Other amendments that can be used to increase the pH of soil include wood ash, industrial CaO (burnt lime), and oyster shells. White firewood ash includes metal salts which are important for processes requiring ions such as Na^+ (sodium), K^+ (potassium), Ca^{2+} (calcium), which may or may not be good for the select flora, but decreases the acidic quality of soil. These products increase the pH of soils through the reaction of CO_3^{2-} with H^+ to produce CO_2 and H_2O. Calcium silicate neutralizes active acidity in the soil by removing free hydrogen ions, thereby increasing pH. As its silicate anion captures H^+ ions (raising the pH), it forms monosilicic acid (H_4SiO_4), a neutral solute.

Decreasing Ph of Alkaline Soil

- Iron sulfates or aluminium sulfate as well as elemental sulfur (S) reduce pH through the formation of sulfuric acid.
- Urea, urea phosphate, ammonium nitrate, ammonium phosphates, ammonium sulfate and monopotassium phosphate fertilizers.
- organic matter in the form of plant litter, compost, and manure will decrease soil pH through the decomposition process. Certain acid organic matter such as pine needles, pine sawdust and acid peat are effective at reducing pH.

SOIL CARBON

Soil carbon includes both inorganic carbon as carbonate minerals, and as soil organic matter. Soil carbon plays a key role in the carbon cycle, and thus it is important in global climate models.

Soil carbon is present in two forms: inorganic and organic. **Soil inorganic carbon** consists of mineral forms of C, either from weathering of parent material, or from reaction of soil minerals with atmospheric CO_2. Carbonate minerals are the dominant form of soil carbon in desert climates. **Soil organic carbon** is present as soil organic matter. It includes relatively available C as fresh plant remains and relatively inert C in materials derived from plant remains: humus and charcoal.

Global Carbon Cycle

Of the 2,700 Gt of C stored in soils worldwide, 1550 GtC is organic and 950 GtC is inorganic carbon, which is approximately three times greater than the current atmospheric C and 240 times higher compared with the current annual fossil fuel emission. The balance of soil carbon is held in peat and wetlands (150 GtC), and in plant litter at the soil surface (50 GtC). This compares to 780 GtC in the atmosphere, and 600 GtC in all living organisms. The oceanic pool holds 38,200 GtC.

About 60 GtC/yr is added to soil. This 60 GtC/yr is the balance of 120 GtC/yr taken out of the atmosphere by terrestrial plant photosynthesis reduced by 60 GtC/yr plant respiration. An equivalent 60 GtC/yr is respired from soil, joining the 60G tC/yr plant respiration to return to the atmosphere.

Organic Carbon

Soil organic carbon is divided between living soil biota and dead biotic material derived from biomass. Together these comprise the soil food web, with the living component sustained by the biotic material component. Soil biota includes earthworms, nematodes, protozoa, fungi, bacteria and different arthropods.

Detritus resulting from plant senescence is the major source of soil organic carbon. Plant materials, with cell walls high in cellulose and lignin, are decomposed and the not-respired carbon is retained as humus. Cellulose and starches are easily degraded, resulting in relatively short residence times. More persistent forms of organic C include lignin, humus, organic matter encapsulated in soil aggregates, and charcoal. These resist alteration and have long residence times.

Soil organic carbon tends to be concentrated in the topsoil. Topsoil ranges from 0.5% to 3.0% organic C for most upland soils. Soils with less than 0.5% organic C are mostly limited to desert areas. Soils containing greater than 12 - 18% organic C are generally classified as organic soils. High levels of organic C develop in soils supporting wetland ecology, flood deposition, fire ecology, and human activity.

Fire derived forms of carbon are present in most soils as unweathered charcoal and weathered black carbon. Soil organic C is typically 5 - 50% derived from char, with levels above 50% encountered in mollisol, chernozem, and terra preta soils.

Root exudates are another source of soil carbon. 5 - 20% of the total plant carbon fixed during photosynthesis is supplied as root exudates in support of rhizospheric mutualistic biota. Microbial populations are typically higher in the rhizosphere than in adjacent bulk soil.

Soil Health

Organic carbon is vital to soil capacity to provide edaphicecosystem services. The condition of this capacity is termed soil health, a term that communicates the value of understanding soil as a living system as opposed to an abiotic component. Specific carbon related benchmarks used to evaluate soil health include CO_2 release, humus levels, and microbial metabolic activity.

A portable soil respiration system measuring soil CO_2 flux

Losses

The exchange of carbon between soils and the atmosphere is a significant part of the world carbon cycle. Carbon, as it relates to the organic matter of soils, is a major component of soil and catchment health. Several factors affect the variation that exists in soil organic matter and soil carbon; the most significant has, in contemporary times, been the influence of humans and agricultural systems.

Although exact quantities are difficult to measure, human activities have caused massive losses of soil organic carbon. First was the use of fire, which removes soil cover and leads to immediate and continuing losses of soil organic carbon. Tillage and drainage both expose soil organic matter to oxygen and oxidation. In the Netherlands, East Anglia, Florida, and the California Delta, subsidence of peat lands from oxidation has been severe as a result of tillage and drainage. Grazing management that exposes soil (through either excessive or insufficient recovery periods) can also cause losses of soil organic.

Managing Soil Carbon

Natural variations in SOM occur as a result of climate, organisms, parent material, time, and relief. The greatest contemporary influence has been that of humans; for example, historical SOM in Australian agricultural soils may have been twice the present range that is typically from 1.6 to 4.6 per cent.

It has long been encouraged that farmers adjust practices to maintain or increase the organic component in the soil. On one hand, practices that hasten oxidation of carbon (such as burning crop stubbles or over-cultivation) are discouraged; on the other hand, incorporation of organic material (such as in manuring) has been encouraged. Increasing soil carbon is not a straightforward matter; it is made complex by the relative activity of soil biota, which can consume and release carbon and are made more active by the addition of nitrogen fertilizers.

Data available on soil Organic Carbon

Europe

The most homogeneous and comprehensive data on the organic carbon/matter content of European soils remain those that can be extracted and/or derived from the European Soil Database in combination with associated databases on land cover, climate, and topography. The modelled data refer to carbon content (%) in the surface horizon of soils in Europe. In an inventory on available national datasets, seven member states of the European Union have available datasets on organic carbon. In the article "Estimating soil organic carbon in Europe based on data collected through an European network" (Ecological Indicators 24, pp. 439–450), a comparison of national data with modelled data is performed. Finally, a new proposed model for estimation of soil organic carbon in agricultural soils has estimated current top SOC stock of 17.63 Gt in EU agricultural soils.

Managing for Catchment Health

Much of the contemporary literature on soil carbon relates to its role, or potential, as an atmospheric carbon sink to offset climate change. Despite this emphasis, a much wider range of soil and catchment health aspects are improved as soil carbon is increased. These benefits are difficult to quantify, due to the complexity of natural resource systems and the interpretation of what constitutes soil health; nonetheless, several benefits are proposed in the following points:

- Reduced erosion, sedimentation: increased soil aggregate stability means greater resistance to erosion; mass movement is less likely when soils are able to retain structural strength under greater moisture levels.
- Greater productivity: healthier and more productive soils can contribute to positive socio-economic circumstances.
- Cleaner waterways, nutrients and turbidity: nutrients and sediment tend to be retained by the soil rather than leach or wash off, and are so kept from waterways.
- Water balance: greater soil water holding capacity reduces overland flow and recharge to groundwater; the water saved and held by the soil remains available for use by plants.
- Climate change: Soils have the ability to retain carbon that may otherwise exist as atmospheric CO_2 and contribute to global warming.
- Greater biodiversity: soil organic matter contributes to the health of soil flora and, accordingly, the natural links with biodiversity in the greater biosphere.

Forest Soils

Forest soils constitute a large pool of carbon. Anthropogenic activities such as deforestation cause releases of carbon from this pool, which may significantly increase the concentration of greenhouse gas (GHG) in the atmosphere. Under the United Nations Framework Convention on Climate Change (UNFCCC), countries must estimate and report GHG emissions and removals, including changes in carbon stocks in all five pools (above- and below-ground biomass, dead wood, litter, and soil carbon) and associated emissions and removals from land use, land-use change and forestry activities, according to the Intergovernmental Panel on Climate Change's good practice guidance. Tropical deforestation represents nearly 25 percent of total anthropogenic GHG emissions worldwide. Deforestation, forest degradation, and changes in land management practices can cause releases of carbon from soil to the atmosphere. For these reasons, reliable estimates of soil organic carbon stock and stock changes are needed for Reducing Emissions from Deforestation and Forest Degradation and GHG reporting under the UNFCCC.

The government of Tanzania—together with the Food and Agriculture Organization of the United Nations and the financial support of the government of Finland—have implemented a forest soil carbon monitoring program to estimate soil carbon stock, using both survey and modelling-based methods.

Nutrients

Sixteen elements or nutrients are essential for plant growth and reproduction. They are carbon C, hydrogen H, oxygen O, nitrogen N, phosphorus P, potassium K, sulfur S, calcium Ca, magnesium Mg, iron Fe, boron B, manganese Mn, copper Cu, zinc Zn, molybdenum Mo, and chlorine Cl. Nutrients required for plants to complete their life cycle are considered essential nutrients. Nutrients that enhance the growth of plants but are not necessary to complete the plant's life cycle are considered non-essential. With the exception of carbon, hydrogen and oxygen, which are supplied by carbon dioxide and water, the nutrients derive originally from the mineral component of the soil.

Plant uptake of nutrients can only proceed when they are present in a plant-available form. In most situations, nutrients are absorbed in an ionic form from (or together with) soil water. Although minerals are the origin of most nutrients, and the bulk of most nutrient elements in the soil is held in crystalline form within primary and secondary minerals, they weather too slowly to support rapid plant growth. For example, The application of finely ground minerals, feldspar and apatite, to soil seldom provides the necessary amounts of potassium and phosphorus at a rate sufficient for good plant growth, as most of the nutrients remain bound in the crystals of those minerals.

The nutrients adsorbed onto the surfaces of clay colloids and soil organic matter provide a more accessible reservoir of many plant nutrients (e.g. K, Ca, Mg, P, Zn). As plants absorb the nutrients from the soil water, the soluble pool is replenished from the surface-bound pool. The decomposition of soil organic matter by microorganisms is another mechanism whereby the soluble pool of nutrients is replenished - this is important for the supply of plant-available N, S, P, and B from soil.

Gram for gram, the capacity of humus to hold nutrients and water is far greater than that of clay minerals. All in all, small amounts of humus may remarkably increase the soil's capacity to promote plant growth.

Plant nutrients, their chemical symbols, and the ionic forms common in soils and available for plant uptake

Element	Symbol	Ion or molecule
Carbon	C	CO_2 (mostly through leaves)
Hydrogen	H	H^+, HOH (water)
Oxygen	O	O^{2-}, OH^-, CO_3^{2-}, SO_4^{2-}, CO_2
Phosphorus	P	$H_2PO_4^-$, HPO_4^{2-} (phosphates)
Potassium	K	K^+
Nitrogen	N	NH_4^+, NO_3^- (ammonium, nitrate)
Sulfur	S	SO_4^{2-}
Calcium	Ca	Ca^{2+}
Iron	Fe	Fe^{2+}, Fe^{3+} (ferrous, ferric)
Magnesium	Mg	Mg^{2+}
Boron	B	H_3BO_3, $H_2BO_3^-$, $B(OH)_4^-$
Manganese	Mn	Mn^{2+}
Copper	Cu	Cu^{2+}
Zinc	Zn	Zn^{2+}
Molybdenum	Mo	MoO_4^{2-} (molybdate)
Chlorine	Cl	Cl^- (chloride)

Soil Processes Important for Nutrient Uptake

Nutrients in the soil are taken up by the plant through its roots. To be taken up by a plant, a nutrient element must be located near the root surface; however, the supply of nutrients in contact with the root is rapidly depleted. There are three basic mechanisms whereby nutrient ions dissolved in the soil solution are brought into contact with plant roots:

- Mass flow of water
- Diffusion within water
- Interception by root growth

All three mechanisms operate simultaneously, but one mechanism or another may be most important for a particular nutrient. For example, in the case of calcium, which is generally plentiful in the soil solution, mass flow alone can usually bring sufficient amounts to the root surface. However, in the case of phosphorus, diffusion is needed to supplement mass flow. For the most part, nutrient ions must travel some distance in the soil solution to reach the root surface. This movement can take place by mass flow, as when dissolved nutrients are carried along with the soil water flowing toward a root that is actively drawing water from the soil. In this type of movement, the nutrient ions are somewhat analogous to leaves floating down a stream. In addition, nutrient ions continually move by diffusion from areas of greater concentration toward the nutrient-depleted areas of lower concentration around the root surface. That process is due to random motion of molecules. By this means, plants can continue to take up nutrients even at night, when water is only slowly absorbed into the roots as transpiration has almost stopped. Finally, root interception comes into play as roots continually grow into new, undepleted soil.

Estimated relative importance of mass flow, diffusion and root interception as mechanisms in supplying plant nutrients to corn plant roots in soils

Nutrient	**Approximate percentage supplied by:**		
	Mass flow	**Root interception**	**Diffusion**
Nitrogen	98.8	1.2	0
Phosphorus	6.3	2.8	90.9
Potassium	20.0	2.3	77.7
Calcium	71.4	28.6	0
Sulfur	95.0	5.0	0
Molybdenum	95.2	4.8	0

In the above table, phosphorus and potassium nutrients move more by diffusion than they do by mass flow in the soil water solution, as they are rapidly taken up by the roots creating a concentration of almost zero near the roots (the plants cannot transpire enough water to draw more of those nutrients near the roots). The very steep concentration gradient is of greater influence in the movement of those ions than is the movement of those by mass flow. The movement by mass flow requires the transpiration of water from the plant causing water and solution ions to also move toward the roots. Movement by root interception is slowest as the plants must extend their roots.

Plants move ions out of their roots in an effort to move nutrients in from the soil. Hydrogen H^+ is exchanged for other cations, and carbonate (HCO_3^-) and hydroxide (OH^-) anions are exchanged for nutrient anions. As plant roots remove nutrients from the soil water solution, they are replenished as other ions move off of clay and humus (by ion exchange or desorption), are added from the weathering of soil minerals, and are released by the decomposition of soil organic matter. Plants derive a large proportion of their anion nutrients from decomposing organic matter, which typically holds about 95 percent of the soil nitrogen, 5 to 60 percent of the soil phosphorus and about 80 percent of the soil sulfur. Where crops are produced, the replenishment of nutrients in the soil must usually be augmented by the addition of fertilizer or organic matter. Because nutrient uptake is an active metabolic process, conditions that inhibit root metabolism may also inhibit nutrient uptake. Examples of such conditions include waterlogging or soil compaction resulting in poor soil aeration, excessively high or low soil temperatures, and above-ground conditions that result in low translocation of sugars to plant roots.

Carbon

Measuring soil respiration in the field using an SRS2000 system.

Plants obtain their carbon from atmospheric carbon dioxide. About 45% of a plant's dry mass is carbon; plant residues typically have a carbon to nitrogen ratio (C/N) of between 13:1 and 100:1. As the soil organic material is digested by arthropods and micro-organisms, the C/N decreases as the carbonaceous material is metabolized and carbon dioxide (CO_2) is released as a byproduct which then finds its way out of the soil and into the atmosphere. The nitrogen is sequestered in the bodies of the living matter of those decomposing organisms and so it builds up in the soil. Normal CO_2 concentration in the atmosphere is 0.03%, this can be the factor limiting plant growth. In a field of maize on a still day during high light conditions in the growing season, the CO_2 concentration drops very low, but under such conditions the crop could use up to 20 times the normal concentration. The respiration of CO_2 by soil micro-organisms decomposing soil organic matter contributes an important amount of CO_2 to the photosynthesising plants. Within the soil, CO_2 concentration is 10 to 100 times that of atmospheric levels but may rise to toxic levels if the soil porosity is low or if diffusion is impeded by flooding.

Nitrogen

Nitrogen is the most critical element obtained by plants from the soil and nitrogen deficiency often limits plant growth. Plants can use the nitrogen as either the ammonium cation (NH_4^+) or the anion nitrate (NO_3^-). Usually, most of the nitrogen in soil is bound within organic compounds that make up the soil organic matter, and must be mineralized to the ammonium or nitrate form before it can be taken up by most plants. The total nitrogen content depends largely on the soil organic matter content, which in turn depends on the climate, vegetation, topography, age and soil management. Soil nitrogen typically decreases by 0.2 to 0.3% for every temperature increase by 10 °C. Usually, grassland soils contain more soil nitrogen than forest soils. Cultivation decreases soil nitrogen by exposing soil organic matter to decomposition by microorganisms, and soils under no-tillage maintain more soil nitrogen than tilled soils.

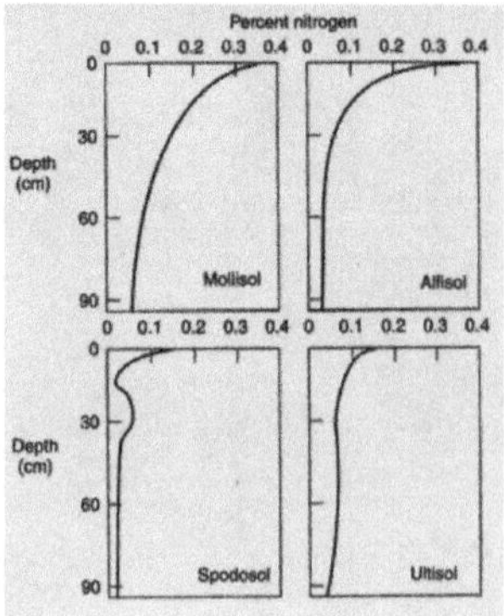

Generalization of percent soil nitrogen by soil order

Some micro-organisms are able to metabolise organic matter and release ammonium in a process called *mineralisation*. Others take free ammonium and oxidise it to nitrate. Nitrogen-fixing bacteria are capable of metabolising N_2 into the form of ammonia in a process called nitrogen fixation. Both ammonium and nitrate can be *immobilized* by their incorporation into the microbes' living cells, where it is temporarily sequestered in the form of amino acids and protein. Nitrate may also be lost from the soil when bacteria metabolise it to the gases N_2 and N_2O. The loss of gaseous forms of nitrogen to the atmosphere due to microbial action is called *denitrification*. Nitrogen may also be *leached* from the soil if it is in the form of nitrate or lost to the atmosphere as ammonia due to a chemical reaction of ammonium with alkaline soil by way of a process called *volatilisation*. Ammonium may also be sequestered in clay by *fixation*. A small amount of nitrogen is added to soil by rainfall.

Nitrogen Gains

In the process of mineralisation, microbes feed on organic matter, releasing ammonia (NH_3), ammonium (NH_4^+) and other nutrients. As long as the carbon to nitrogen ratio (C/N) of fresh residues in the soil is above 30:1, nitrogen will be in short supply and other bacteria will feed on the ammonium and incorporate its nitrogen into their cells in the immobilization process. In that form the nitrogen is said to be *immobilised*. Later, when such bacteria die, they too are *mineralised* and some of the nitrogen is released as ammonium and nitrate. If the C/N is less than 15, ammonia is freed to the soil, where it may be used by bacteria which oxidise it to nitrate (nitrification). Bacteria may on average add 25 pounds (11 kg) nitrogen per acre, and in an unfertilised field, this is the most important source of usable nitrogen. In a soil with 5% organic matter perhaps 2 to 5% of that is released to the soil by such decomposition. It occurs fastest in warm, moist, well aerated soil. The mineralisation of 3% of the organic material of a soil that is

4% organic matter overall, would release 120 pounds (54 kg) of nitrogen as ammonium per acre.

Carbon/Nitrogen Ratio of Various Organic Materials

Organic Material	C:N Ratio
Alfalfa	13
Bacteria	4
Clover, green sweet	16
Clover, mature sweet	23
Fungi	9
Forest litter	30
Humus in warm cultivated soils	11
Legume-grass hay	25
Legumes (alfalfa or clover), mature	20
Manure, cow	18
Manure, horse	16–45
Manure, human	10
Oat straw	80
Straw, cornstalks	90
Sawdust	250

In nitrogen fixation, rhizobium bacteria convert N_2 to ammonia (NH_3). Rhizobia share a symbiotic relationship with host plants, since rhizobia supply the host with nitrogen and the host provides rhizobia with nutrients and a safe environment. It is estimated that such symbiotic bacteria in the root nodules of legumes add 45 to 250 pounds of nitrogen per acre per year, which may be sufficient for the crop. Other, free-living nitrogen-fixing bacteria and blue-green algae live independently in the soil and release nitrate when their dead bodies are converted by way of mineralisation.

Some amount of usable nitrogen is fixed by lightning as nitric oxide (NO) and nitrogen dioxide (NO_2^-). Nitrogen dioxide is soluble in water to form nitric acid (HNO_3) solution of H^+ and NO_3^-. Ammonia, NH_3, previously released from the soil or from combustion, may fall with precipitation as nitric acid at a rate of about five pounds nitrogen per acre per year.

Nitrogen Sequestration

When bacteria feed on soluble forms of nitrogen (ammonium and nitrate), they temporarily sequester that nitrogen in their bodies in a process called *immobilisation*. At a later time when those bacteria die, their nitrogen may be released as ammonium by the processes of mineralisation.

Protein material is easily broken down, but the rate of its decomposition is slowed by its attachment to the crystalline structure of clay and when trapped between the clay layers. The layers are small enough that bacteria cannot enter. Some organisms can exude extracellular enzymes that can act on the sequestered proteins. However, those enzymes too may be trapped on the clay crystals.

Ammonium fixation occurs when ammonium pushes potassium ions from between the layers of clay such as illite or montmorillonite. Only a small fraction of soil nitrogen is held this way.

Nitrogen Losses

Usable nitrogen may be lost from soils when it is in the form of nitrate, as it is easily leached. Further losses of nitrogen occur by denitrification, the process whereby soil bacteria convert nitrate (NO_3^-) to nitrogen gas, N_2 or N_2O. This occurs when poor soil aeration limits free oxygen, forcing bacteria to use the oxygen in nitrate for their respiratory process. Denitrification increases when oxidisable organic material is available and when soils are warm and slightly acidic. Denitrification may vary throughout a soil as the aeration varies from place to placc. Denitrification may cause the loss of 10 to 20 percent of the available nitrates within a day and when conditions are favourable to that process, losses of up to 60 percent of nitrate applied as fertiliser may occur.

Ammonium volatilisation occurs when ammonium reacts chemically with an alkaline soil, converting NH_4^+ to NH_3. The application of ammonium fertiliser to such a field can result in volatilisation losses of as much as 30 percent.

Phosphorus

After nitrogen, phosphorus is probably the element most likely to be deficient in soils. The soil mineral apatite is the most common mineral source of phosphorus. While there is on average 1000 lb of phosphorus per acre in the soil, it is generally unavailable in the form of phosphates of low solubility. Total phosphorus is about 0.1 percent by weight of the soil, but only one percent of that is available. Of the part available, more than half comes from the mineralisation of organic matter. Agricultural fields may need to be fertilised to make up for the phosphorus that has been removed in the crop.

When phosphorus does form solubilised ions of $H_2PO_4^-$, they rapidly form insoluble phosphates of calcium or hydrous oxides of iron and aluminum. Phosphorus is largely immobile in the soil and is not leached but actually builds up in the surface layer if not cropped. The application of soluble fertilisers to soils may result in zinc deficiencies as zinc phosphates form. Conversely, the application of zinc to soils may immobilise phosphorus again as zinc phosphate. Lack of phosphorus may interfere with the normal opening of the plant leaf stomata, resulting in plant temperatures 10 percent higher than normal. Phosphorus is most available when soil pH is 6.5 in mineral soils and 5.5 in organic soils.

Potassium

The amount of potassium in a soil may be as much as 80,000 lb per acre-foot, of which only 150 lb is available for plant growth. Common mineral sources of potassium are the mica biotite and potassium feldspar, $KAlSi_3O_8$. When solubilised, half will be held as exchangeable cations on clay while the other half is in the soil water solution. Potassium fixation often occurs when soils dry and the potassium is bonded between layers of illite clay. Under certain conditions, dependent on the soil texture, intensity of drying, and initial amount of exchangeable potassium, the fixed percentage may be as much as 90 percent within ten minutes. Potassium may be leached from soils low in clay.

Calcium

Calcium is one percent by weight of soils and is generally available but may be low as it is soluble and can be leached. It is thus low in sandy and heavily leached soil or strongly acidic mineral soil. Calcium is supplied to the plant in the form of exchangeable ions and moderately soluble minerals. Calcium is more available on the soil colloids than is potassium because the common mineral calcite, $CaCO_3$, is more soluble than potassium-bearing minerals.

Magnesium

Magnesium is one of the dominant exchangeable cations in most soils (as are calcium and potassium). Primary minerals that weather to release magnesium include hornblende, biotite and vermiculite. Soil magnesium concentrations are generally sufficient for optimal plant growth, but highly weathered and sandy soils may be magnesium deficient due to leaching by heavy precipitation.

Sulfur

Most sulfur is made available to plants, like phosphorus, by its release from decomposing organic matter. Deficiencies may exist in some soils (especially sandy soils) and if cropped, sulfur needs to be added. The application of large quantities of nitrogen to fields that have marginal amounts of sulfur may cause sulfur deficiency in the rapidly growing plants by the plant's growth outpacing the supply of sulfur. A 15-ton crop of onions uses up to 19 lb of sulfur and 4 tons of alfalfa uses 15 lb per acre. Sulfur abundance varies with depth. In a sample of soils in Ohio, United States, the sulfur abundance varied with depths, 0-6 inches, 6-12 inches, 12-18 inches, 18-24 inches in the amounts: 1056, 830, 686, 528 lb per acre respectively.

Micronutrients

The micronutrients essential for plant life, in their order of importance, include iron, manganese, zinc, copper, boron, chlorine and molybdenum. The term refers to plants' needs, not to their abundance in soil. They are required in very small amounts but are essential to plant health in that most are required parts of some enzyme system which speeds up plants' metabolisms. They are generally available in the mineral component of the soil, but the heavy application of phosphates can cause a deficiency in zinc and iron by the formation of insoluble zinc and iron phosphates. Iron deficiency may also result from excessive amounts of heavy metals or calcium minerals (lime) in the soil. Excess amounts of soluble boron, molybdenum and chloride are toxic.

Non-Essential Nutrients

Nutrients which enhance the health but whose deficiency does not stop the life cycle of plants include: cobalt, strontium, vanadium, silicon and nickel.As their importance are evaluated they may be added to the list of essential plant nutrients.

SOIL GAS

Soil gases are the gases found in the air space between soil components. The primary natural soil gases include nitrogen, carbon dioxide and oxygen. The oxygen is critical because it allows for respiration of both plant roots and soil organisms. Other natural soil gases are atmospheric methane and radon. Some environmental contaminants below ground produce gas which diffuses through the soil such as from landfill wastes, mining activities, and contamination by petroleum hydrocarbons which produce volatile organic compounds. Soil gases can diffuse into buildings, the chief concerns among these pollutants are radon which is radioactive and causes cancer and methane which can be flammable at only 4.4% concentration.

Gases fill soil pores in the soil structure as water drains or is removed from a soil pore by evaporation or root absorption. The network of pores within the soil aerates, or ventilates, the soil. This aeration network becomes blocked when water enters soil pores. Not only are both soil air and soil water very dynamic parts of soil, but both are often inversely related.

Composition of air in soil and atmosphere:

- Nitrogen: Soil Air: 79.2% Atmosphere: 79.0%
- Oxygen: Soil Air: 20.6% Atmosphere: 20.9%
- Carbon Dioxide: Soil Air: 0.25% Atmosphere: 0.04%

Gas molecules in soil are in continuous thermal motion according to the kinetic theory of gases, there is also collision between molecules - a random walk.

In soil, a concentration gradient causes net movement of molecules from high concentration to low concentration, this gives the movement of gas by diffusion. Numerically, it is explained by Fick›s law of diffusion.

Soil Atmosphere

The atmosphere of soil is radically different from the atmosphere above. The consumption of oxygen, by microbes and plant roots and their release of carbon dioxide, decrease oxygen and increase carbon dioxide concentration. Atmospheric CO_2 concentration is 0.04%, but in the soil pore space it may range from 10 to 100 times that level. At extreme levels CO_2 is toxic. In addition, the soil voids are saturated with water vapour. Adequate porosity is necessary, not just to allow the penetration of water, but also to allow gases to diffuse in and out. Movement of gases is by diffusion from high concentrations to lower. Oxygen diffuses in and is consumed and excess levels of carbon dioxide, diffuse out with other gases as well as water. Soil texture and structure strongly affect soil porosity and gas diffusion. It is the total pore space (porosity) of soil not the pore size that determines the rate of diffusion of gases into and out of soil. A Platy soil structure and compacted soils (low porosity) impede gas flow, and a deficiency of oxygen may encourage anaerobic bacteria to reduce (strip oxygen) from nitrate NO_3 to the gases N_2, N_2O, and NO, which are then lost to the atmosphere, thereby depleting the soil of nitrogen. Aerated soil is also a net sink of methane CH_4 but a net producer of greenhouse gases when soils are depleted of oxygen and subject to elevated temperatures.

REFERENCES

1. Wezel, A., Bellon, S., Doré, T., Francis, C., Vallod, D., David, C. (2009). Agroecology as a science, a movement or a practice. A review. Agronomy for Sustainable Development (published online)
2. Wibblemann et al. (2013) Mainstreaming Agroecology: Implications for Global Food and Farming Systems. http://futureoffood.org/pdfs/Coventry_University_2013_Maintstreaming_Agroecology.pdf
3. Pretty, Jules. 2008. Agricultural sustainability: concepts, principles and evidence. Philosophical Transactions of the Royal Society, 363, 447-465.
4. Conway, Gordon R. 1985. Agroecosystem analysis. Agricultural Administration, 20, 31-55.
5. Agroecology, Glossary of Statistical Terms
6. Dalgaard, Tommy, and Nicholas Hutchings, John Porter. Agroecology, Scaling and Interdisciplinarity. Agriculture Ecosystems and Environment 100(2003): 39-51.
7. *Francis; et al. (2003). "Agroecology: the ecology of food systems". Journal of Sustainable Agriculture.* ***22*** *(3): 99–118. doi:10.1300/J064v22n03_10.*
8. Agroecology.org
9. Buttel, Frederick. Envisioning the Future Development of Farming in the USA: Agroecology between Extinction and Multifunctionality? New Directions in Agroecology Research and Education (2003)
10. Wojtkowski, Paul A. (2002) Agroecological Perspectives in Agronomy, Forestry and Agroforestry. Science Publishers Inc., Enfield, NH, 356p.
11. Wojtkowski, Paul A. (2006) Introduction to Agroecology: Principles and Practices.

Haworth Press, Binghamton, NY, 404p.

12. IFOAM (International Federation for Organic Agriculture Movements)
13. Boer, I J. M. 2003. Environmental impact assessment of conventional and organic milk production. Livestock Production Science. Vol 80, p 69–77.
14. Hovi, M. el al. 2003. Animal health and welfare in organic livestock production in Europe: current state and future challenges. Livestock Production Science. Vol 80, p 41–53.
15. Bennedsgaard, T.W. et al. 2003. Eleven years of organic dairy production in Denmark: herd health and production related to time of conversion and compared to conventional production. Livestock production science. Vol 80, p 121-131.
16. Garcia-Torres, L. et al. 2002. Summary of the Workshop on Soil Protection and Sustainable Agriculture organized by the EU Commission DG Environment and the DG Environmental Quality of the Spanish Ministry of Environment (Soria, Spain)
17. Branco, H. and Lal, R. 2008. Principles of Conservation Management. No Tillage-Farming (Ch.8). Springer Verlag. Netherlands. P. 195
18. Bolliger, A. et al. 2006. Taking stock of the Brazilian Zero Till Revolution: A review of landmark research and farmers› practice. Adv. Agron. Vol 91, p 47–110
19. Calegari, A. et al. 2008. Impact of Long-Term No-Tillage and Cropping System Management on Soil Organic Carbon in an Oxisol: A Model for Sustainability. Agronomy Journal. Vol 100, Issue 4, p 1013-1019
20. West, T. and Post, W. 2002. Soil Organic Carbon Sequestration Rates by Tillage and Crop Rotation: A Global Data Analysis. Soil Sci. Soc. Am. J. 66:1930–1946
21. Machado, P.L.O.A. and Silva, C.A. 2001. Soil management under no-tillage systems in the tropics with special reference to Brazil. Nutr. Cycling Agroecosyst. Vol 61, p 119–130
22. Koga, N. et al. 2003. Fuel consumption-derived CO_2 emissions under conventional and reduced tillage cropping systems in northern Japan. Agriculture, Ecosystems and Environment. Vol 99, p 213–219.
23. Fleizo et al. 2002
24. Koller, K. 2003. Techniques of soil tillage. Ed. Adel El Titi (CRC Press)
25. Pavuk, D.M. 1994. Influence of weeds within Zea mays crop plantings on populations of adult Diabrotica barberi and Diabrotica virgifera virgifera. Agriculture, Ecosystems & Environment. Vol 50, p 165-175
26. G. E. Brust, B. Stinner & D. McCartney. 1986. Predator activity and predation in corn agroecosystems. Environmental Entomology 15:1017-1021
27. Randall, G.W., and P.R. Hill. 2000. Fall strip-tillage systems. p. 193–199. In R.C. Reeder (ed.) Conservation tillage systems and management. MWPS-45, 2nd ed. Iowa State Univ., Ames.
28. Licht, M.A. and Al-Kaisi, M. 2005. Strip-tillage effect on seedbed soil temperature and other soil physical properties. Soil and Tillage Research. Vol 80, p 233-249

29. Gliessman, Stephen. R Agroecology: Ecological Processes in Sustainable Agriculture. Ann Arbor: Sleeping Bear Press, 1998.
30. Klages, K.H.W. 1928. Crop ecology and ecological crop geography in the agronomic curriculum. J. Amer. Soc. Agron. 20:336-353.
31. Wezel, A., Soldat, V. (2009). A quantitative and qualitative historical analysis of the scientific discipline agroecology. International Journal of Agricultural Sustainability 7 (1): 3-18.
32. Tischler, W. (1965). Agrarökologie. Gustav Fischer Verlag, Jena, Germany, 499 pp.
33. Friederichs, K. (1930) Die Grundfragen und Gesetzmäßigkeiten der land- und forstwirtschaftlichen Zoologie. Vol. 1: Ökologischer Teil, Vol. 2: Wirtschaftlicher Teil. Verlagsbuchhandlung Paul Parey, Berlin, Germany, 417 and 443 pp.
34. Hansen, B., Alrøe, H.F., Kristensen, E.S., 2001. Approaches to assess the environmental impact of organic farming with particular regard to Denmark. Agric. Ecosys. Environ. 83, 11–26.
35. Harper, J.L., 1974. Agric. Ecosyst. Agroecosyst. 1, 1–6.
36. qtd. in Francis *et al.* 2003.
37. Reproduced from Francis et al., 2003 and Wezel et al., 2009.

CHAPTER 5

Soil Salinity

Chapter Contents

- **The Soil Salinity Problem**
- **Principles of Salinity Control**
- **Soil Leaching**
- **Strip Cropping: An Alternative**
- **Soil Salinity Models**
- **Erosion Control**
- **Soil Value**

Soil salinity control relates to controlling the problem of soil salinity and reclaiming salinized agricultural land.

The aim of soil salinity control is to prevent soil degradation by salination and reclaim already salty (saline) soils. Soil reclamation is also called soil improvement, rehabilitation, remediation, recuperation, or amelioration.

The primary man-made cause of salinization is irrigation. River water or groundwater used in irrigation contains salts, which remain behind in the soil after the water has evaporated.

The primary method of controlling soil salinity is to permit 10-20% of the irrigation water to leach the soil, be drained and discharged through an appropriate drainage system. The salt concentration of the drainage water is normally 5 to 10 times higher than that of the irrigation water, thus salt export matches salt import and it will not accumulate.

THE SOIL SALINITY PROBLEM

Salty (saline) soils are soils that have a high salt content. The predominant salt is normally sodium chloride (NaCl, "table salt"). Saline soils are therefore also *sodic soils* but there may be sodic soils that are not saline, but alkaline.

World Soil Salt Degradation

This damage is an average of 2,000 hectares of irrigated land in arid and semi-arid areas daily for more than 20 years across 75 countries (each week the world loses an area larger than Manhattan)...To feed the world's anticipated nine billion people by 2050, and with little new productive land available, it's a case of all lands needed on deck.—*principal author Manzoor Qadir, Assistant Director, Water and Human Development, at UN University's Canadian-based Institute for Water, Environment and Health*

According to a study by UN University, about 62 million hectares (20%) of the world's irrigated lands are affected, up from 45 million hectares in the early 1990s. In the Indo-Gangetic Plain, home to over 10% of the world's population, crop yield losses for wheat, rice, sugarcane and cotton grown on salt-affected lands could be 40%, 45%, 48%, and 63%, respectively.

Salty soils are a common feature and an environmental problem in irrigated lands in arid and semi-arid regions, resulting in poor or little crop production. The problems are often associated with high water tables, caused by a lack of natural subsurface drainage to the underground. Poor subsurface drainage may be caused by insufficient transport capacity of the aquifer or because water cannot exit the aquifer, for instance if the aquifer is situated in a topographical depression.

Worldwide, the major factor in the development of saline soils is a lack of precipitation. Most naturally saline soils are found in (semi)arid regions and climates of the earth.

Primary Cause

For example, assuming irrigation water with a low salt concentration of 0.3 g/l (equal to 0.3 kg/m^3 corresponding to an electric conductivity of about 0.5 FdS/m) and a modest annual supply of irrigation water of 10,000 m^3/ha (almost 3 mm/day) brings 3,000 kg salt/ha each year. In the absence of sufficient natural drainage (as in waterlogged soils) and without a proper leaching and drainage program to remove salts, this would lead to a high soil salinity and reduced crop yields in the long run.

Irrigated saline land with poor crop stand

Much of the water used in irrigation has a higher salt content than in this example, which is compounded by the fact that many irrigation projects use a far greater annual supply of water. Sugar cane, for example, needs about 20,000 m^3/ha of water per year. As a result, irrigated areas often receive more than 3,000 kg/ha of salt per year and some receive as much as 10,000 kg/ha/year.

Secondary Cause

The secondary cause of salinization is waterlogging in irrigated land. Irrigation causes changes to the natural water balance of irrigated lands. Large quantities of water in irrigation projects are not consumed by plants and must go somewhere. In irrigation projects it is impossible to achieve 100% irrigation efficiency where all the irrigation water is consumed by the plants. The maximum attainable irrigation efficiency is about 70% but usually it is less than 60%. This means that minimum 30%, but usually more than 40% of the irrigation water is not evaporated and it must go somewhere.

Most of the water lost this way is stored underground which can change the original hydrology of local aquifers considerably. Many aquifers cannot absorb and transport these quantities of water and so the water table rises leading to water logging.

Waterlogging causes three problems:

- The shallow water table and lack of oxygenation of the root zone reduces the yield of most crops
- It leads to an accumulation of salts brought in with the irrigation water as their removal through the aquifer is blocked
- With the upward seepage of groundwater more salts are brought into the soil and the salination is aggravated

Aquifer conditions in irrigated land and the groundwater flow have an important role in soil salinization, as illustrated here :

Illustration of the influence of aquifer conditions on soil salinization in irrigated land

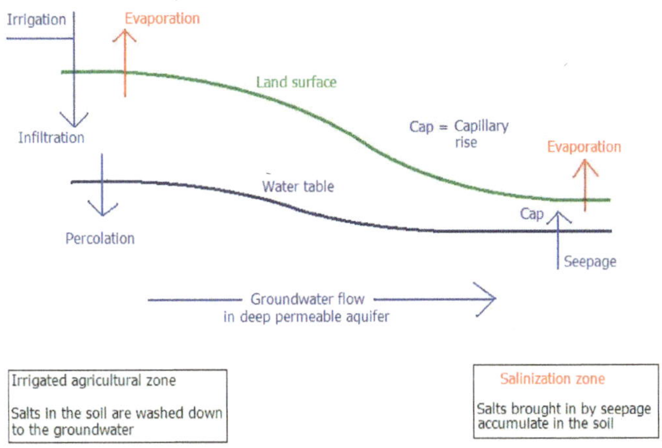

Soil salinization in the lower parts of undulating land with a good aquifer

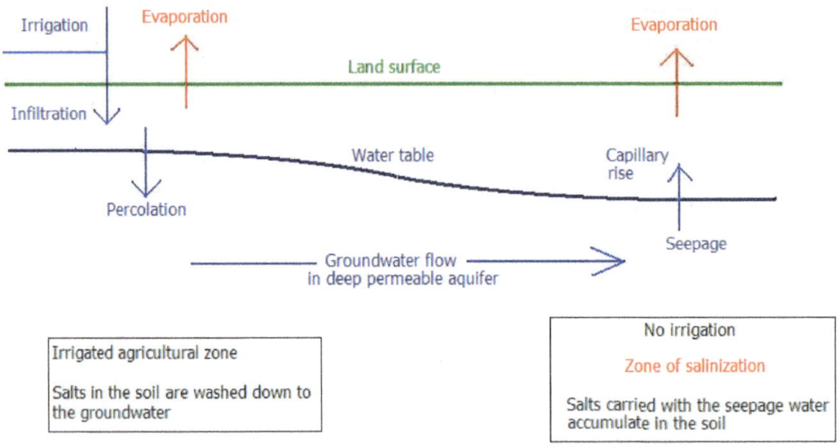

Soil salinization in the unirrigated parts of flat land with a good aquifer

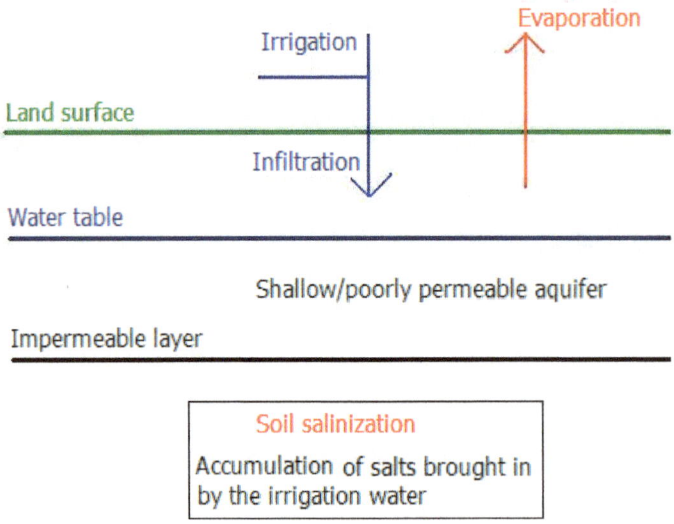

Soil salinization in irrigated flat land without an aquifer

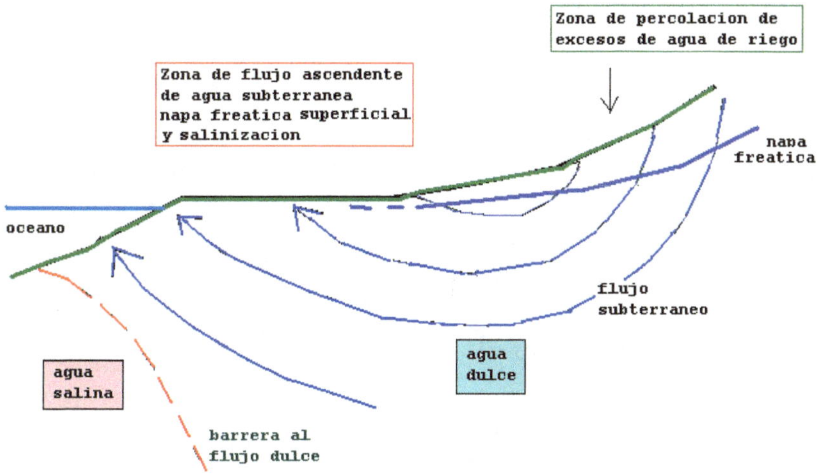

Soil salinization in a coastal delta from irrigation higher up

Salt Affected Area

Normally, the salinization of agricultural land affects a considerable area of irrigation projects, on the order of 20 to 30%. When the agriculture in such a fraction of the land is abandoned, a new salt and water balance is attained, a new equilibrium is reached, and the situation becomes stable.

In India alone, thousands of square kilometres have been severely salinized. China and Pakistan do not lag much behind (perhaps China has even more salt affected land than India). A regional distribution of the 3,230,000 km^2 of saline land worldwide is shown in the following table derived from the FAO/UNESCO Soil Map of the World.

Region	Area (10^{6}ha)
Australia	84.7
Africa	69.5
Latin America	59.4
Near and Middle East	53.1
Europe	20.7
Asia and Far East	19.5
Northern America	16.0

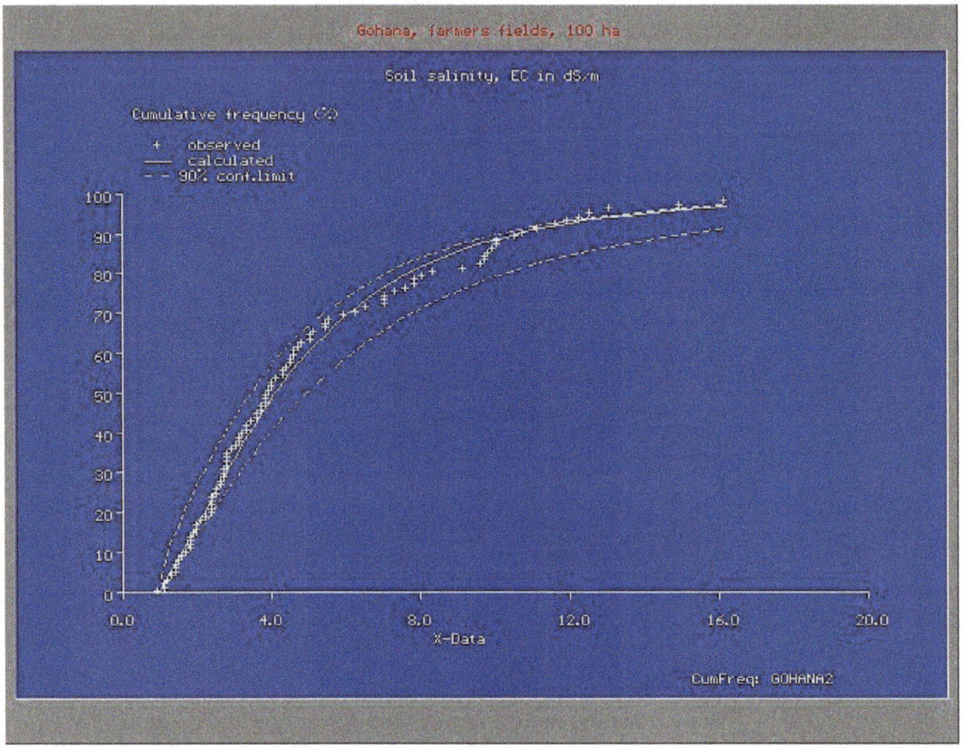

CumFreq program: spatial variation of soil salinity

Spatial Variation

Although the principles of the processes of salinization are fairly easy to understand, it is more difficult to explain why certain parts of the land suffer from the problems and other parts do not, or to predict accurately which part of the land will fall victim. The main reason for this is the variation of natural conditions in time and space, the usually uneven distribution of the irrigation water, and the seasonal or yearly changes of agricultural practices. Only in lands with undulating topography is the prediction simple: the depressional areas will degrade the most.

The preparation of salt and water balances for distinguishable sub-areas in the irrigation project, or the use of agro-hydro-salinity models, can be helpful in explaining or predicting the extent and severity of the problems.

DIAGNOSIS

Measurement

Soil salinity is measured as the salt concentration of the soil solution in tems of g/l or electric conductivity (EC) in dS/m. The relation between these two units is about

5/3 : y g/l => 5y/3 dS/m. Seawater may have a salt concentration of 30 g/l (3%) and an EC of 50 dS/m.

The standard for the determination of soil salinity is from an extract of a saturated paste of the soil, and the EC is then written as ECe. The extract is obtained by centrifugation. The salinity can more easily be measured, without centrifugation, in a 2:1 or 5:1 water:soil mixture (in terms of g water per g dry soil) than from a saturated paste. The relation between ECe and $EC_{2:1}$ is about 4, hence : ECe = 4 $EC_{1:2}$.

Classification

Soils are considered saline when the ECe > 4. When 4 < ECe < 8, the soil is called slightly saline, when 8 < ECe < 16 it is called (moderately) saline, and when ECe > 16 severely saline.

Crop Tolerance

Sensitive crops lose their vigor already in slightly saline soils, most crops are negatively affected by (moderately) saline soils, and only salinity resistant crops thrive in severely saline soils. The University of Wyoming and the Government of Alberta report data on the salt tolerance of plants.

PRINCIPLES OF SALINITY CONTROL

Drainage is the primary method of controlling soil salinity. The system should permit a small fraction of the irrigation water (about 10 to 20 percent, the drainage or leaching fraction) to be drained and discharged out of the irrigation project.

In irrigated areas where salinity is stable, the salt concentration of the drainage water is normally 5 to 10 times higher than that of the irrigation water. Salt export matches salt import and salt will not accumulate.

When reclaiming already salinized soils, the salt concentration of the drainage water will initially be much higher than that of the irrigation water (for example 50 times higher). Salt export will greatly exceed salt import, so that with the same drainage fraction a rapid desalinization occurs. After one or two years, the soil salinity is decreased so much, that the salinity of the drainage water has come down to a normal value and a new, favorable, equilibrium is reached.

In regions with pronounced dry and wet seasons, the drainage system may be operated in the wet season only, and closed during the dry season. This practice of checked or controlled drainage saves irrigation water.

The discharge of salty drainage water may pose *environmental problems* to downstream areas. The environmental hazards must be considered very carefully and, if necessary mitigating measures must be taken. If possible, the drainage must be limited to wet seasons only, when the salty effluent inflicts the least harm.

Drainage Systems

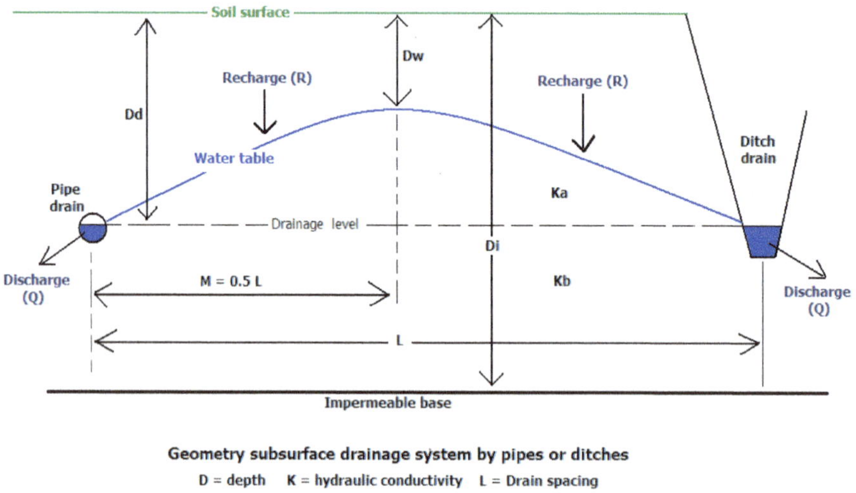

Parameters of a horizontal drainage system

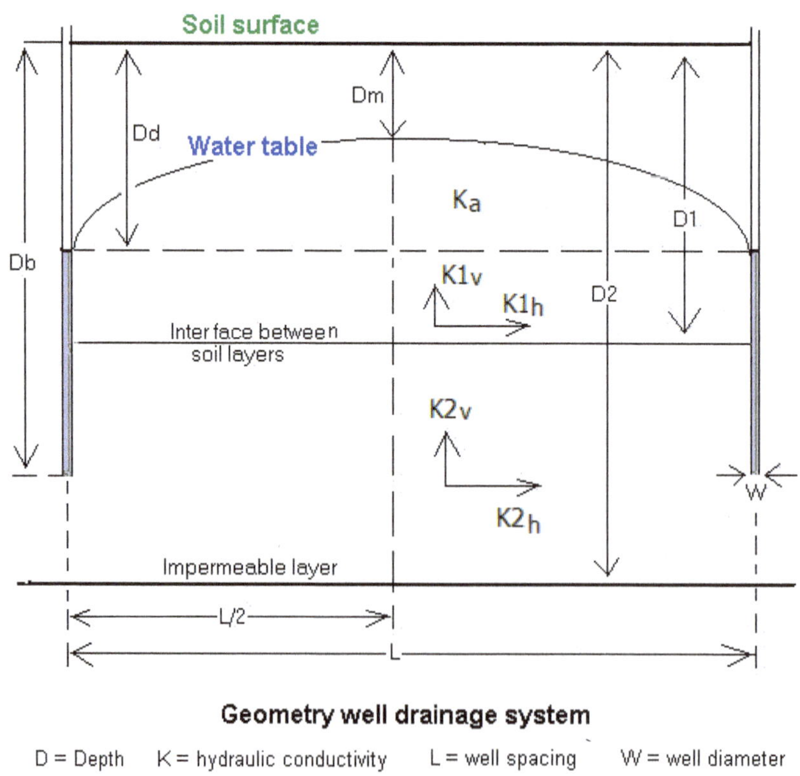

Parameters of a vertical drainage system

Land drainage for soil salinity control is usually by *horizontal drainage system* (figure left), but *vertical systems* (figure right) are also employed.

The drainage system designed to evacuate salty water also lowers the water table. To reduce the cost of the system, the lowering must be reduced to a minimum. The highest permissible level of the water table (or the shallowest permissible depth) depends on the irrigation and agricultural practices and kind of crops.

In many cases a seasonal average water table depth of 0.6 to 0.8 m is deep enough. This means that the water table may occasionally be less than 0.6 m (say 0.2 m just after an irrigation or a rain storm). This automatically implies that, in other occasions, the water table will be deeper than 0.8 m (say 1.2 m). The fluctuation of the water table helps in the breathing function of the soil while the expulsion of carbon dioxide (CO_2) produced by the plant roots and the inhalation of fresh oxygen (O_2) is promoted.

The establishing of a not too deep water table offers the additional advantage that excessive field irrigation is discouraged, as the crop yield would be negatively affected by the resulting elevated water table, and irrigation water may be saved.

The statements made above on the optimum depth of the water table are very general, because in some instances the required water table may be still shallower than indicated (for example in rice paddies), while in other instances it must be considerably deeper (for example in some orchards). The establishment of the optimum depth of the water table is in the realm of agricultural drainage criteria.

SOIL LEACHING

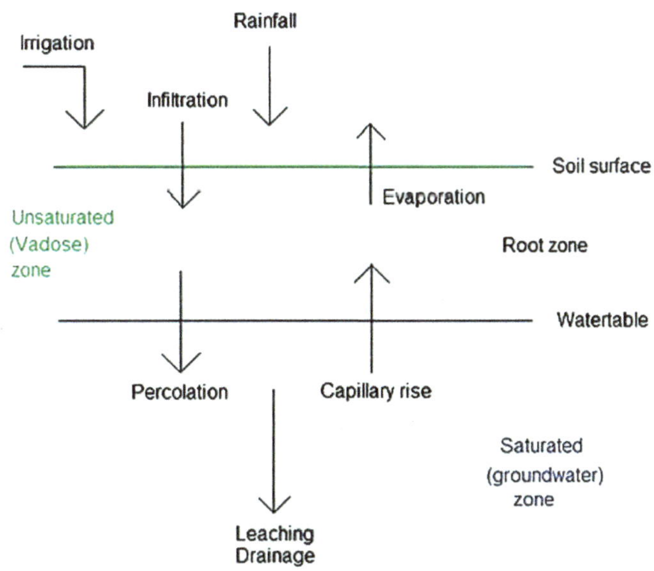

Water balance factors in the soil

The vadose zone of the soil below the soil surface and the watertable is subject to four main hydrological inflow and outflow factors:

- Infiltration of rain and irrigation water (Irr) into the soil through the soil surface (Inf) :
- Inf = Rain + Irr
- Evaporation of soil water through plants and directly into the air through the soil surface (Evap)
- Percolation of water from the unsaturated zone soil into the groundwater through the watertable (Perc)
- Capillary rise of groundwater moving by capillary suction forces into the unsaturated zone(Cap)

In steady state (i.e. the amount of water stored in the unsaturated zone does not change in the long run) the water balance of the unsaturated zone reads: Inflow = Outflow, thus:

- Inf + Cap = Evap + Perc or :
- Irr + Rain + Cap = Evap + Perc

and the *salt balance* is

- Irr.Ci + Cap.Cc = Evap.Fc.Ce + Perc.Cp + Ss

where Ci is the salt concentration of the irrigation water, Cc is the salt concentration of the capillary rise, equal to the salt concentration of the upper part of the groundwater body, Fc is the fraction of the total evaporation transpired by plants, Ce is the salt concentration of the water taken up by the plant roots, Cp is the salt concentration of the percolation water, and Ss is the increase of salt storage in the unsaturated soil. This assumes that the rainfall contains no salts. Only along the coast this may not be true. Further it is assumed that no runoff or surface drainage occurs. The amount of removed by plants (Evap.Fc.Ce) is usually negligibly small: Evap.Fc.Ce = 0

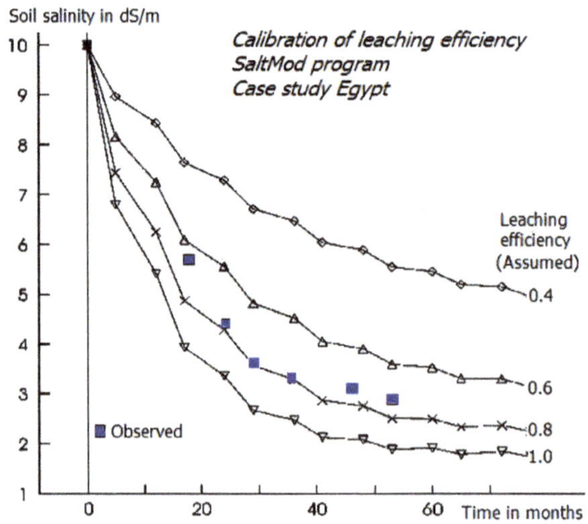

Leaching curves, calibrating leaching efficiency

The salt concentration Cp can be taken as a part of the salt concentration of the soil in the unsaturated zone (Cu) giving: Cp=Le.Cu, where Le is the leaching efficiency. The leaching efficiency is often in the order of 0.7 to 0.8, but in poorly structured, heavy clay soils it may be less. In the Leziria Grande polder in the delta of the Tagus river in Portugal it was found that the leaching efficiency was only 0.15.Assuming that one wishes to avoid the soil salinity to increase and maintain the soil salinity Cu at a desired level Cd we have: Ss = 0, Cu = Cd and Cp = Le.Cd. Hence the salt balance can be simplified to:

- Perc.Le.Cd = Irr.Ci + Cap.Cc

Setting the amount percolation water required to fulfill this salt balance equal to Lr (the *leaching requirement*) it is found that:

- Lr = (Irr.Ci + Cap.Cc) / Le.Cd .

Substituting herein Irr = Evap + Perc − Rain − Cap and re-arranging gives :

- Lr = [(Evap−Rain).Ci + Cap(Cc−Ci)] / (Le.Cd − Ci)

With this the irrigation and drainage requirements for salinity control can be computed too. In irrigation projects in (semi)arid zones and climates it is important to check the leaching requirement, whereby the *field irrigation efficiency* (indicating the fraction of irrigation water percolating to the underground) is to be taken into account. The desired soil salinity level Cd depends on the crop tolerance to salt. The University of Wyoming, USA, and the Government of Alberta, Canada, report crop tolerance data.

STRIP CROPPING: AN ALTERNATIVE

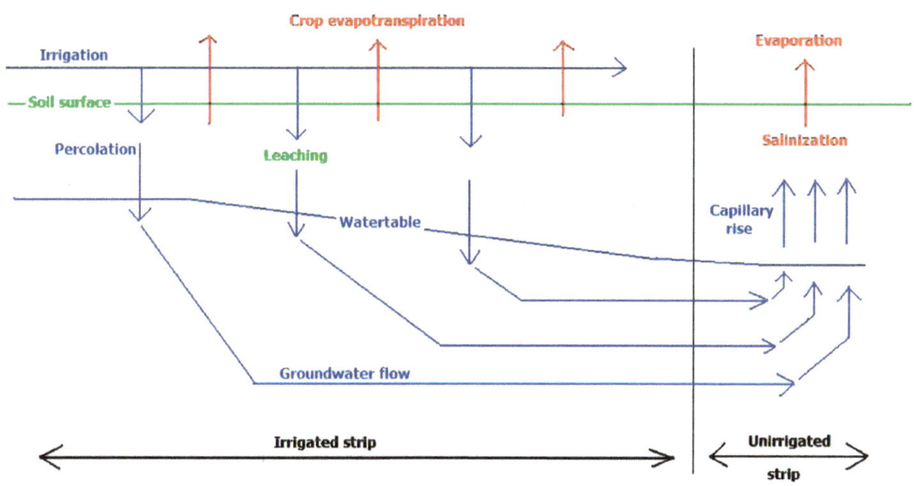

Principle of strip cropping and land sacrifice

Hydrological principles of *strip cropping* to control the depth of the water table and the soil salinity

In irrigated lands with scarce water resources suffering from drainage (high water table) and soil salinity problems, strip cropping is sometimes practiced with strips of land where every other strip is irrigated while the strips in between are left permanently fallow.

Owing to the water application in the irrigated strips they have a higher watertable which induces flow of groundwater to the unirrigated strips. This flow functions as subsurface drainage for the irrigated strips, whereby the water table is maintained at a not-too-shallow depth, leaching of the soil is possible, and the soil salinity can be controlled at an acceptably low level.

In the unirrigated (sacrificial) strips the soil is dry and the groundwater comes up by capillary rise and evaporates leaving the salts behind, so that here the soil salinizes. Nevertheless, they can have some use for livestock, sowing salinity resistant grasses or weeds. Moreover, useful salt resistant trees can be planted like Casuarina, Eucalyptus or Atriplex, keeping in mind that the trees have deep rooting systems and the salinity of the wet subsoil is less than of the topsoil. In these ways wind erosion can be controlled. The unirrigated strips can also be used for salt harvesting.

SOIL SALINITY MODELS

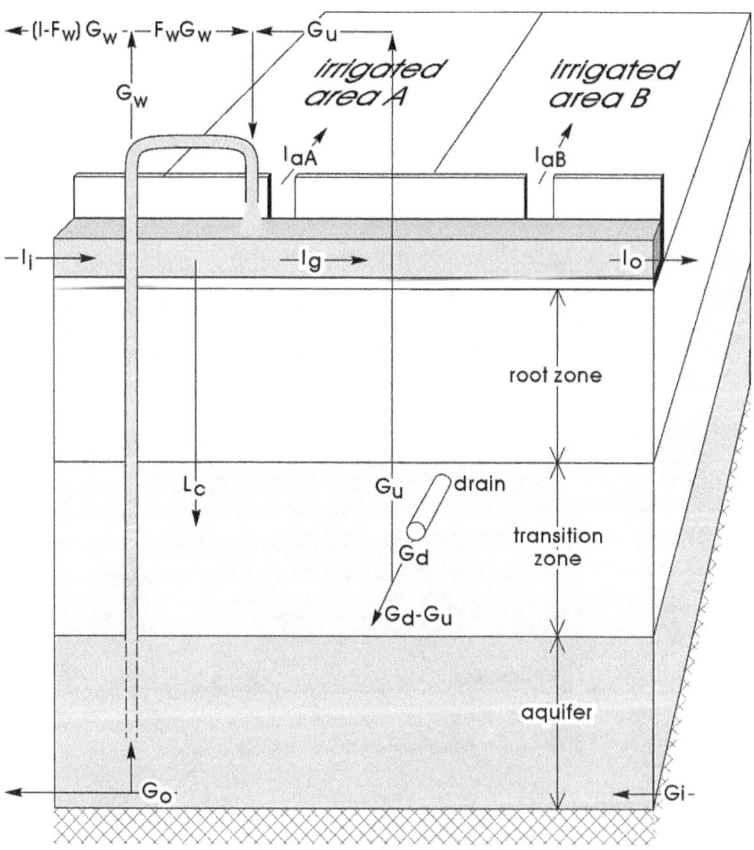

SaltMod components

The majority of the computer models available for water and solute transport in the soil (e.g. SWAP, DrainMod-S,UnSatChem, and Hydrus) are based on Richard's differential equation for the movement of water in unsaturated soil in combination with Fick's differential convection–diffusion equation for advection and dispersion of salts.

The models require input of soil characteristics like the relations between variable unsaturated soil moisture content, water tension, water retention curve, unsaturated hydraulic conductivity, dispersivity and diffusivity. These relations vary to a great extent from place to place and from time to time and are not easy to measure. Further, the models are difficult to calibrate under farmer's field conditions because the soil salinity here is spatially very variable. The models use short time steps and need at least a daily, if not an hourly, data base of hydrological phenomena. Altogether this makes model application to a fairly large project the job of a team of specialists with ample facilities.

Simpler models, like SaltMod, based on monthly or seasonal water and soil balances and an empirical capillary rise function, are also available. They are useful for long-term salinity predictions in relation to irrigation and drainage practices.

LeachMod, using the SaltMod principles, helps in analyzing leaching experiments in which the soil salinity was monitored in various root zone layers while the model will optimize the value of the leaching efficiency of each layer so that a fit is obtained of observed with simulated soil salinity values.

Spatial variations owing to variations in topography can be simulated and predicted using salinity cum groundwater models, like SahysMod.

EROSION CONTROL

Terraces, conservation tillage, and conservation buffers save soil and improve water quality on this Iowa farm.

Hydroseeding in UK

Erosion control is the practice of preventing or controlling wind or water erosion in agriculture, land development, coastal areas, river banks and construction. Effective erosion controls handle surface runoff and are important techniques in preventing water pollution, soil loss, wildlife habitat loss and human property loss.

Usage

Erosion controls are used in natural areas, agricultural settings or urban environments. In urban areas erosion controls are often part of stormwater runoff management programs required by local governments. The controls often involve the creation of a physical barrier, such as vegetation or rock, to absorb some of the energy of the wind or water that is causing the erosion. They also involve building and maintaining storm drains. On construction sites they are often implemented in conjunction with sediment controls such as sediment basins and silt fences.

Bank erosion is a natural process: without it, rivers would not meander and change course. However, land management patterns that change the hydrograph and/or vegetation cover can act to increase or decrease channel migration rates. In many places, whether or not the banks are unstable due to human activities, people try to keep a river in a single place. This can be done for environmental reclamation or to prevent a river from changing course into land that is being used by people. One way that this is done is by placing riprap or gabions along the bank.

Examples

Examples of erosion control methods include:

- buffer strip
- cellular confinement systems
- crop rotation
- conservation tillage
- contour bunding
- contour plowing
- cover crops
- fiber rolls

- gabions
- hydroseeding
- level spreaders
- mulching
- perennial crops
- plasticulture
- polyacrylamide (as a coagulant)
- reforestation
- riparian strip
- riprap
- strip farming
- sand fence
- vegetated waterway (bioswale)
- terracing
- wattle (construction)
- windbreaks

Mathematical modeling

Since the 1920s and 1930s scientists have been creating mathematical models for understanding the mechanisms of soil erosion and resulting sediment surface runoff, including an early paper by Albert Einstein applying Baer's law. These models have addressed both gully and sheet erosion. Earliest models were a simple set of linked equations which could be employed by manual calculation. By the 1970s the models had expanded to complex computer models addressing nonpoint source pollution with thousands of lines of computer code. The more complex models were able to address nuances in micrometerology, soil particle size distributions and micro-terrain variation.

SOIL VALUE

Soil value (German: *Bodenwertzahl*) or **BWZ** is a comparative assessment of soil quality used in Germany. It is determined from soil sampling data and ranges from 0 (very low) to 100 (very high).

According to the Geological Service of the state of North Rhine-Westphalia, the soil value of arable land is based on a field assessment framework which combines an assessment of soil type, soil condition (soil development) and the parent material of which the soil is composed. The result, the soil value, expresses the relative net income that, under normal and proper management, is determined only by the profitability of the soil. Additions or deductions in this value to take account of variations in yield due to terrain and climate (e.g. average annual temperature) give the field value (*Ackerzahl*).

Range of Values

The following range of values and colour scheme is used:

Very poor	0 - 18	brown
Poor	18 - 35	red
Medium	35 - 55	orange
High	55 - 75	yellow
Very High	75 - 100	green

The values exhibited by different types of soil are:

Sand	0 - 11
Sandy loam	11 - 30
Heavy to clayey loam	31 - 50
Loam, partly with loess covering	51 - 70
Loam with loess covering	71 - 90
Loess	91 -100

Origins

As a yardstick for soil quality, a "Reich Standard Farm" was set up during the Third Reich in the parish of Eickendorf in the Magdeburger Börde, an exceptionally rich arable region. During the Reich Soil Assessment (*Reichsbodenschätzung*) following the Soil Assessment Act in 1934 a soil value of 100 was established in the leading Reich farm of *Haberhauffe/Jäger*. It was the basis of comparison for the tax rating of farms in Germany. Since this farm was no longer available for comparisons within West Germany after the Second World War, a farm in Machtsum near Harsum in the Hildesheim Börde was designated as the Federal Standard Farm. During later measurements, an even higher value of BWZ - 102.8 - was measured in Mölme, about 20 km east of Hildesheim within the municipality of Söhlde. It is the highest value ever recorded in Germany.

Another measure, the field value (*Ackerzahl*) is derived from the soil value, taking account of other factors such as climate and location e.g. along the edge of woodland.

REFERENCES

1. http://phys.org/news/2014-10-world-hectares-farm-soil-daily.html
2. I.P. Abrol, J.S.P Yadav, and F. Massoud 1988. Salt affected soils and their management, Food and Agricultural Organization of the United Nations (FAO), Soils Bulletin 39.
3. ILRI, 2003. *Drainage for Agriculture: Drainage and hydrology/salinity - water and salt balances*. Lecture notes International Course on Land Drainage, International Institute for Land Reclamation and Improvement (ILRI), Wageningen, The Netherlands. Download from page : , or directly as PDF :

4. R.Brinkman, 1980. Saline and sodic soils. In: Land reclamation and water management, p. 62-68. International Institute for Land Reclamation and Improvement (ILRI), Wageningen, The Netherlands.

5. SaltMod: *a tool for interweaving of irrigation and drainage for salinity control*. In: W.B.Snellen (ed.), Towards integration of irrigation, and drainage management. ILRI Special report, p. 41-43. International Institute for Land Reclamation and Improvement (ILRI), Wageningen, The Netherlands.

6. ILRI, 2003, *This paper discusses soil salinity*. Lecture notes, International Course on Land Drainage International Institute for Land Reclamation and Improvement (ILRI), Wageningen, The Netherlands. On line:

7. L.A.Richards (Ed.), 1954. Diagnosis and improvement of saline and alkali soils. USDA Agricultural Handbook 60. On internet

8. Alan D. Blaylock, 1994, *Soil Salinity and Salt tolerance of Horticultural and Landscape Plants.*

9. Government of Alberta, *Salt tolerance of Plants*

10. J.W. van Hoorn and J.G. van Alphen (2006), Salinity control. In: H.P. Ritzema (Ed.), Drainage Principles and Applications, p. 533-600, Publication 16, International Institute for Land Reclamation and Improvement (ILRI), Wageningen, The Netherlands. ISBN 90-70754-33-9.

11. Agricultural Drainage Criteria, Chapter 17 in: H.P.Ritzema (2006), Drainage Principles and Applications, Publication 16, International Institute for Land Reclamation and Improvement (ILRI), Wageningen, The Netherlands. ISBN 90-70754-33-9. On line :

12. R.J.Oosterbaan and M.A.Senna, 1990. Using SaltMod to predict drainage and salinity control in the Nile delta. In: Annual Report 1989, International Institute for Land Reclamation and Improvement (ILRI), Wageningen, The Netherlands, p. 63-74. See *Case Study Egypt* in the SaltMod manual :

13. E.A. Vanegas Chacon, 1990. Using SaltMod to predict desalinization in the Leziria Grande Polder, Portugal. Thesis. Wageningen Agricultural University, The Netherlands

14. ILRI, 2000. *Irrigation, groundwater, drainage and soil salinity control in the alluvial fan of Garmsar*. Consultancy assignment to the Food and Agriculture Organization (FAO) of the United Nations, International Institute for Land Reclamation and Improvement (ILRI), Wageningen, The Netherlands. On line:

15. SWAP model

16. DrainMod-S model

17. UnSatChem model

18. Hydrus model

19. LeachMod

CHAPTER 6

Soil Classification

Chapter Contents

Soil classification is the separation of soil into classes or groups each having similar characteristics and potentially similar behaviour. A classification for engineering purposes should be based mainly on mechanical properties, e.g. permeability, stiffness, strength.

INTRODUCTION

Alfisols are a soil order in USDA soil taxonomy. Alfisols form in semiarid to humid areas, typically under a hardwood forest cover. They have a clay-enriched subsoil and relatively high native fertility. "Alf" refers to aluminium (Al) and iron (Fe). Because of their productivity and abundance, the Alfisols represent one of the more important soil orders for food and fiber production. They are widely used both in agriculture and forestry, and are generally easier to keep fertile than other humid-climate soils, though those in Australia and Africa are still very deficient in nitrogen and available phosphorus. Those in monsoonal tropical regions, however, have a tendency to acidify when heavily cultivated, especially when nitrogenous fertilizers are used.

In the FAO soil classification, most Alfisols are classified as Luvisols or Lixisols, but some are classed as Nitosols.

Alfisols occupy around one-tenth of the Earth's ice-free land surface. They are dominant in many areas, such as the Ohio River basin in the United States, southern and unglaciated

Western Europe, the Baltic region and central European Russia, the drier parts of Peninsular India, Sudan in Africa, and many parts of South America.

Alfisols have undergone only moderate leaching. By definition, they have at least 35% base saturation, meaning calcium, magnesium, and potassium are relatively abundant. This is in contrast to Ultisols, which are the more highly leached forest soils having less than 35% base saturation. In eastern North America, Alfisols are commonly found in glaciated areas while Ultisols are restricted to the areas south of the limit of maximum glaciation.

The fossil record of Alfisols begins in the Late Devonian. Probably owing to their fertility, they are the oldest forest soils; vegetation on weathered Oxisols, by contrast, is not known earlier than Middle Permian. Fossil Alfisols remain common from the Carboniferous and all periods since the Eocene.

An Alfisol profile.

Suborders

- Aqualfs - This suborder is saturated with water long enough to cause oxygen depletion. Almost all of this suborder is thought to have been forested at some point. Redoximorphic features are present.
- Cryalfs - This suborder occurs mainly in high elevations which tend to be cold with temperatures ranging from greater than 0 C to less than 8 C.
- Udalfs - This suborder is located in humid or subhumid climates. All of this suborder is thought to have been forested at some point.
- Ustalfs - This suborder is located in semiarid climates with wet summers and dry winters.

- Xeralfs - This suborder is located in mediterranean climates with very dry summers and wet winters.

Mapping

Most empirical knowledge of soil in nature comes from soil survey efforts. Soil survey, or soil mapping, is the process of determining the soil types or other properties of the soil cover over a landscape, and mapping them for others to understand and use. It relies heavily on distinguishing the individual influences of the five classic soil forming factors. This effort draws upon geomorphology, physical geography, and analysis of vegetation and land-use patterns. Primary data for the soil survey are acquired by field sampling and supported by remote sensing.

Soil Classification Systems

Australia

There are fourteen soil orders at the top level of the Australian Soil Classification. They are: Anthroposols, Organosols, Podosols, Vertosols, Hydrosols, Kurosols, Sodosols, Chromosols, Calcarosols, Ferrosols, Dermosols, Kandosols, Rudosols and Tenosols.

European Union

The EU's soil taxonomy is based on a new standard soil classification in the World Reference Base for Soil Resources produced by the UN's Food and Agriculture Organization. According to this, the major soils in the European Union are:

- Acrisols
- Albeluvisols
- Andosols
- Anthrosols
- Arenosols
- Calcisols
- Cambisols
- Chernozems
- Fluvisols
- Gleysols
- Gypsisols
- Histosols
- Kastanozems
- Leptosols
- Luvisols

- Phaeozems
- Planosols
- Podzols
- Regosols
- Solonchaks
- Solonetzes
- Umbrisols
- Vertisols

USA

A taxonomy is an arrangement in a systematic manner; the USDA soil taxonomy has six levels of classification. They are, from most general to specific: order, suborder, great group, subgroup, family and series. Soil properties that can be measured quantitatively are used in this classification system — they include: depth, moisture, temperature, texture, structure, cation exchange capacity, base saturation, clay mineralogy, organic matter content and salt content. There are 12 soil orders (the top hierarchical level) in the USDA soil taxonomy. The names of the orders end with the suffix *-sol*. The criteria for the different soil orders include properties that reflect major differences in the genesis of soils. The orders are:

- Alfisol - soils with aluminium and iron. They have horizons of clay accumulation, and form where there is enough moisture and warmth for at least three months of plant growth. They constitute 10.1% of soils worldwide.
- Andisol - volcanic ash soils. They are young and very fertile. They cover 1% of the world's ice-free surface.
- Aridisol - dry soils forming under desert conditions which have fewer than 90 consecutive days of moisture during the growing season and are nonleached. They include nearly 12% of soils on Earth. Soil formation is slow, and accumulated organic matter is scarce. They may have subsurface zones of caliche or duripan. Many aridisols have well-developed Bt horizons showing clay movement from past periods of greater moisture.
- Entisol - recently formed soils that lack well-developed horizons. Commonly found on unconsolidated river and beach sediments of sand and clay or volcanic ash, some have an A horizon on top of bedrock. They are 18% of soils worldwide.
- Gelisol - permafrost soils with permafrost within two metres of the surface or gelic materials and permafrost within one metre. They constitute 9.1% of soils worldwide.
- Histosol - organic soils, formerly called bog soils, are 1.2% of soils worldwide.
- Inceptisol - young soils. They have subsurface horizon formation but show little eluviation and illuviation. They constitute 15% of soils worldwide.

- Mollisol - soft, deep, dark fertile soil formed in grasslands and some hardwood forests with very thick A horizons. They are 7% of soils worldwide.
- Oxisol - are heavily weathered, are rich in iron and aluminum oxides (sesquioxides) or kaolin but low in silica. They have only trace nutrients due to heavy tropical rainfall and high temperatures and low CEC of the remaining clays. They are 7.5% of soils worldwide.
- Spodosol - acid soils with organic colloid layer complexed with iron and aluminium leached from a layer above. They are typical soils of coniferous and deciduous forests in cooler climates. They constitute 4% of soils worldwide.
- Ultisol - acid soils in the humid tropics and subtropics, which are depleted in calcium, magnesium and potassium (important plant nutrients). They are highly weathered, but not as weathered as Oxisols. They make up 8% of the soil worldwide.
- Vertisol - inverted soils. They are clay-rich and tend to swell when wet and shrink upon drying, often forming deep cracks into which surface layers can fall. They are difficult to farm or to construct roads and buildings due to their high expansion rate. They constitute 2.4% of soils worldwide.

Alfisol.

Andisol.

Aridisol.

Entisol.

Gelisol.

Histisol.

Inceptisol.

Mollisol.

Oxisol.

Spodosol.

Utisol.

Vertisol.

The percentages listed above are for land area free of ice. "Soils of Mountains", which constitute the balance (11.6%), have a mixture of those listed above, or are classified as "Rugged Mountains" which have no soil.

The above soil orders in sequence of increasing degree of development are Entisols, Inceptisols, Aridisols, Mollisols, Alfisols, Spodosols, Ultisols, and Oxisols. Histosols and Vertisols may appear in any of the above at any time during their development.

The soil suborders within an order are differentiated on the basis of soil properties and horizons which depend on soil moisture and temperature. Forty-seven suborders are recognized in the United States.

The soil great group category is a subdivision of a suborder in which the kind and sequence of soil horizons distinguish one soil from another. About 185 great groups are recognized in the United States. Horizons marked by clay, iron, humus and hard pans and soil features such as the expansion-contraction of clays (that produce self-mixing provided by clay), temperature, and marked quantities of various salts are used as distinguishing features.

The great group categories are divided into three kinds of soil subgroups: typic, intergrade and extragrade. A typic subgroup represents the basic or ‹typical› concept of the great group to which the described subgroup belongs. An intergrade subgroup describes the properties that suggest how it grades towards (is similar to) soils of other soil great groups, suborders or orders. These properties are not developed or expressed well enough to cause the soil to be included within the great group towards which they grade, but suggest similarities. Extragrade features are aberrant properties which prevent that soil from being included in another soil classification. About 1,000 soil subgroups are defined in the United States. A soil family category is a group of soils within a subgroup and describes the physical and chemical properties which affect the response of soil to agricultural management and engineering applications. The principal characteristics used to differentiate soil families include texture, mineralogy, pH, permeability, structure, consistency, the locale's precipitation pattern, and soil temperature. For some soils the criteria also specify the percentage of silt, sand and coarse fragments such as gravel, cobbles and rocks. About 4,500 soil families are recognised in the United States.

A family may contain several soil series which describe the physical location using the name of a prominent physical feature such as a river or town near where the soil sample was taken. An example would be Merrimac for the Merrimack River in New Hampshire, USA. More than 14,000 soil series are recognised in the United States. This permits very specific descriptions of soils.

A soil phase of series, originally called ‹soil type› describes the soil surface texture, slope, stoniness, saltiness, erosion, and other conditions.

Classification

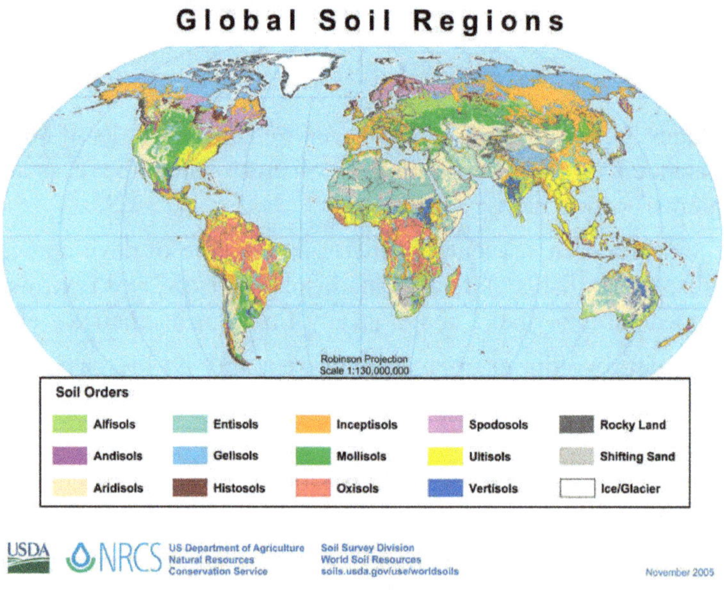

Map of global soil regions from the USDA.

As of 2006, the World Reference Base for Soil Resources, via its Land & Water Development division, is the pre-eminent soil classification system. It replaces the previous FAO soil classification.

The WRB borrows from modern soil classification concepts, including USDA soil taxonomy. The classification is based mainly on soil morphology as an expression pedogenesis. A major difference with USDA soil taxonomy is that soil climate is not part of the system, except insofar as climate influences soil profile characteristics. Many other classification schemes exist, including vernacular systems. The structure in vernacular systems are either nominal, giving unique names to soils or landscapes, or descriptive, naming soils by their characteristics such as red, hot, fat, or sandy. Soils are distinguished by obvious characteristics, such as physical appearance (e.g., color, texture, landscape position), performance (e.g., production capability, flooding), and accompanying vegetation. A vernacular distinction familiar to many is classifying texture as heavy or light. Light soil content and better structure, take less effort to turn and cultivate. Contrary to popular belief, light soils do not weigh less than heavy soils on an air dry basis nor do they have more porosity.

Vasily Dokuchaev, a Russian geologist, geographer and early soil scientist, is credited with identifying soil as a resource whose distinctness and complexity deserved to be separated conceptually from geology and crop production and treated as a whole.

Previously, soil had been considered a product of chemical transformations of rocks, a dead substrate from which plants derive nutritious elements. Soil and bedrock were in fact equated. Dokuchaev considers the soil as a natural body having its own genesis and its own history of development, a body with complex and multiform processes taking place within it. The soil is considered as different from bedrock. The latter becomes soil under the influence of a series of soil-formation factors (climate, vegetation, country, relief and age). According to him, soil should be called the daily or outward horizons of rocks regardless of the type; they are changed naturally by the common effect of water, air and various kinds of living and dead organisms.

A 1914 encyclopedic definition: the different forms of earth on the surface of the rocks, formed by the breaking down or weathering of rocks. serves to illustrate the historic view of soil which persisted from the 19th century. Dokuchaev›s late 19th century soil concept developed in the 20th century to one of soil as earthy material that has been altered by living processes. A corollary concept is that soil without a living component is simply a part of earth›s outer layer.

Further refinement of the soil concept is occurring in view of an appreciation of energy transport and transformation within soil. The term is popularly applied to the material on the surface of the Earth›s moon and Mars, a usage acceptable within a portion of the scientific community. Accurate to this modern understanding of soil is Nikiforoff›s 1959 definition of soil as the excited skin of the sub aerial part of the earth›s crust.

USDA SOIL TAXONOMY

USDA Soil Taxonomy developed by United States Department of Agriculture and the National Cooperative Soil Survey provides an elaborate classification of soil types according to several parameters (most commonly their properties) and in several levels: *Order*, *Suborder*, *Great Group*, *Subgroup*, *Family*, and *Series*.

Soil Orders

Name of Soil Orders in Soil Taxonomy with Their Major Characteristics

Name	Major Characteristics	Name	Major Characteristics
Alfisols	Must have argillic, natric, or kandic horizon; High-to-medium base saturation; Moderately weathered; Commonly form under boreal or broadleaf forests; Rich in iron and aluminum; Common in humid areas, semi-tropics, and mediterranean climates; 9.6% of global & 14.5% of U.S. ice-free land	Andisols	Form from volcanic ejecta, dominated by allophane or Al-humic complexes; Must have andic soil properties: high in poorly crystalline Fe and Al minerals, high in phosphorus, low bulk density, and high proportions of glass and amorphous colloidal materials, such as allophane, imogolite and ferrihydrite; High Organic Matter content, sometimes melanic epipedon; 0.7% of global & 1.7% of U.S. ice-free land
Aridisols	Dry soil (i.e., must have aridic moisture regime); Ochric epipedon is common; Sometimes argillic or natric horizon; Must have some diagnostic subsurface horizon; Commonly in deserts; 12.7% of global & 8.8% of U.S. ice-free land	Entisols	Least soil profile development; Ochric epipedon is common; No B horizons; most common order by surface area (16.3% of global & 12.2% of U.S. ice-free land)
Gelisols	Soils with permafrost within 100 cm or cryoturbation (frost churning) within 100 cm plus permafrost within 200 cm; Commonly at high latitudes and elevations; 8.6% of global & 7.5% of U.S. ice-free land	Histosols	Must have histic epipedon; Usually aquic soil moisture regime; No diagnostic subsurface horizons; Rapid decomposition when aerated; Peat or bog; >20% organic matter; Organic soil materials extending down to an impermeable layer or with an organic layer that is more than 40 cm thick and without andic properties Commonly in wetlands (swamps, marshes, etc.); 1.2% of global & 1.3% of U.S. ice-free land

Inceptisols	Similar to Entisol, but beginning of a B horizon is evident; No diagnostic subsurface horizons; On landscapes continuously eroded or young deposits; Cambic, sulfuric, calcic, gypsic, petrocalcic, or petrogypsic horizon, or with a mollic, umbric, or histic epipedon, or with an exchangeable sodium percentage of >15% or fragipan; 9.9% of global & 9.1% of U.S. ice-free land	Mollisols	Must have mollic epipedon; High base saturation of >50%; Dark soils; Some with argillic or natric horizons; Common in grasslands; 6.9% of global & 22.4% of U.S. ice-free land
Oxisols	Most soil profile development; Must have oxic horizon within 150 cm of soil surfacc; Low nutrient availability; No argillic horizon; Highly weathered; Dominated by end-member clays, Al and Fe oxides; Commonly in old landscapes in tropics; 7.6% of global & <0.01% of U.S. ice-free land	Spodosols	Must have spodic horizon within 2 m of soil surface and without andic properties; Usually have albic horizon; High in Fe, Al oxides and humus accumulation; Acidic soils; Common in coniferous or boreal forests; 2.6% of global and 3.3% of U.S. ice-free land
Ultisols	Must have argillic or kandic horizon; Low base saturation of <35% at 2 m depth or 75 cm below a fragipan; Common in subtropical regions; often known as red clay soils; 8.5% of global & 9.6% of U.S. ice-free land	Vertisols	Usually mollic epipedon; High in shrinking and swelling clays; >30% clay to a depth of 50 cm; Deep cracks (called gilgai) form when soil dries; Form from parent material high in clay (e.g., shales, basins, exposed Bt horizons of old soils); 2.4% of global & 1.7% of U.S. ice-free land

Andisol

In USDA soil taxonomy, **andisols** are soils formed in volcanic ash and defined as soils containing high proportions of glass and amorphous colloidal materials, including allophane, imogolite and ferrihydrite. In the FAO soil classification, andisols are known as **andosols**.

Because they are generally quite young, andisols typically are very fertile except in cases where phosphorus is easily fixed (this sometimes occurs in the tropics). They can usually support intensive cropping, with areas used for wet rice in Java supporting some of the densest populations in the world. Other andisol areas support crops of fruit, maize, tea, coffee or tobacco. In the Pacific Northwest USA, andisols support very productive forests.

Andisols occupy about 1% of the global ice-free land area. Most occur around the Pacific Ring of Fire, with the largest areas found in central Chile, Ecuador, Colombia, Mexico, the Pacific Northwest USA, Japan, Java and New Zealand's North Island. Other areas occur in the Great Rift Valley, Kenya, Italy, Iceland and Hawai.

Fossil andisols are known from areas far from present-day volcanic activity and have in some cases been dated as far back as the Precambrian 1.5 billion years ago.

Suborders

- **Aquands** - Andisols with a water table at or near the surface for much of the year.
- **Gelands** - Andisols of very cold climates (mean annual temperature <0°C).
- **Cryands** - Andisols of cold climates.
- **Torrands** - Andisols of very dry climates.
- **Ustands** - Andisols of semiarid and sub humid climates.
- **Udands** - Andisols of humid climates.
- **Xerands** - temperate Andisols with very dry summers and moist winters.
- **Vitrands** - relatively young Andisols that are coarse-textured and dominated by glass.

Aridisol

Aridisols (or **desert soils**) are a soil order in USA soil taxonomy. Aridisols (from the Latin *aridus*, for "dry", and *solum*) form in an arid or semi-arid climate. Aridisols dominate the deserts and xeric shrublands, which occupy about one third of the Earth's land surface. Aridisols have a very low concentration of organic matter, reflecting the paucity of vegetative production on these dry soils. Water deficiency is the major defining characteristic of aridisols. Also required is sufficient age to exhibit sub-soil weathering and development. Limited leaching in aridisols often results in one or more subsurface soil horizons in which suspended or dissolved minerals have been deposited: silicate clays, sodium, calcium carbonate, gypsum or soluble salts. These subsoil horizons can also be cemented by carbonates, gypsum or silica. Accumulation of salts on the surface can result in salinization.

Entisol

In USDA soil taxonomy, **entisols** are defined as soils that do not show any profile development other than an A horizon. An entisol has no diagnostic horizons, and most are basically unaltered from their parent material, which can be unconsolidated sediment or rock. Entisols are the second most abundant soil order (after inceptisols), occupying about 16% of the global ice-free land area.

In Australia, most entisols are known as *rudosols* or *tenosols*, whilst arents are known as *anthroposols*. In the FAO soil classification, because of the diversity of their properties, suborders of entisols form individual soil orders (e.g. *fluvisols*, *lithosols*).

Entisol profile showing little or no evidence of pedogenic horizon development

Causes of Delayed or Absent Development

- Unweatherable parent materials - sand, iron oxide, aluminium oxide, kaolinite clay.
- Erosion - common on shoulder slopes; other kinds also important.
- Deposition - continuous, repeated deposition of new parent materials by flood as diluvium, aeolian processes which means by wind, slope processes as colluvium, mudflows, other means.
- Flooding or saturation.
- Cold climate - must not be sufficiently cold in winter for permafrost.
- Dry climate.
- Shallow to bedrock - may be rock resistant to weathering, such as quartzite or ironstone.
- Toxic parent materials - serpentine soil, mine spoils, sulfidic clays.

Suborders

- **Aquents** - permanently or usually wet soils formed on river banks, tidal mudflats etc. Here, general wetness limits development.
- **Arents** - anthropogenic soils: diagnostic horizons cannot develop because of deep mixing through plowing, spading, or other methods of moving by humans.

In other soil classifications these are known as anthrosols.

- **Fluvents** - alluvial soils where development is prevented by repeated deposition of sediment in periodic floods. Found in valleys and deltas of rivers, especially those with high sediment load.
- **Orthents** - shallow or "skeletal soils". Found on recent erosional surfaces or very old landforms *completely devoid of weatherable minerals.*
- **Psamments** - entisols that are sandy in all layers where development is precluded by the impossibility of weathering the sand. Formed from shifting or glacialsand dunes.

Paleopedology

Most fossil soils before the development of terrestrial vegetation in the Silurian are entisols, showing no distinct soil horizons. Entisols have been abundant in the paleopedological record ever since then, though, unlike other soil orders (oxisols, ultisols, gelisols for instance) they do not have value as indicators of climate - though orthents might in some cases be indicated of an extremely old landscape with very little soil formation (as in Australia today).

Gelisol

Gelisols are an order in USDA soil taxonomy. They are soils of very cold climates which are defined as containing permafrost within two meters of the soil surface. The word "gelisol comes from the Latin *gelare* meaning "to freeze", a reference to the process of cryoturbation that occurs from the alternating thawing and freezing characteristic of gelisols.

In United Nations Food and Agriculture Organization soil classification system, gelisols are known as **cryosols**.

Structurally, gelisols have no B horizon and have an A horizon resting on the permafrost. Because soil organic matter accumulates in the upper layer, most gelisols are black or dark brown in soil color, followed by a shallow mineral layer. Despite the influence of glaciation in most areas where gelisols occur, chemically they are not highly fertile because nutrients, especially calcium and potassium, are very easily leached above the permafrost. The permafrost greatly restricts the engineering use of gelisols, as large structures (e.g. buildings) subside as the frozen earth thaws when they are put in place.

Gelisols are found chiefly in Siberia, Alaska and Canada. Smaller areas are found in the Andes (mainly near the intersection between Chile, Bolivia and Argentina), Tibet, northern Scandinavia and the ice-free parts of Greenland and Antarctica. Fossil gelisols are known from as far back as Precambrian ice ages 900 million years ago.

Suborders

In USDA soil taxonomy, gelisols are subdivided into:

- **Histels**: organic soils similar to histosols except that they have permafrost within two meters below ground surface. They have 80% or more organic materials from the soil surface to a depth of 50 cm or to a glacic layer or densic, lithic, or paralithic contact, whichever is shallowest. These soils occur predominantely in subarctic and low arctic regions of continuous or widespread permafrost. Less than one-third of the active layer (the soil between the ground surface and a permafrost table) or an ice layer which is at least 30-cm thick has been cryoturbated.

- **Turbels**: soils that show marked influence of cryoturbation (more than one-third of the depth of the active layer) such as irregular, broken, or distorted horizon boundaries and involutions and areas with patterned ground. They commonly contain tongues of mineral and organic horizons, organic and mineral intrusions and oriented rock fragments. Organic matter is accumulated on top of the permafrost and ice wedges are a common features. Turbels occur primarily in the zone of continuous permafrost.
- **Orthels**: soils that show little or no cryoturbation (less than one-third of the depth of the active layer). Patterned ground (except for polygons) generally is lacking. Orthels occur primarily within the zone of discontinuous permafrost, and in alpine areas.

Histosol

In both the FAO soil classification and the USDA soil taxonomy, a **histosol** is a soil consisting primarily of organic materials. They are defined as having 40 centimetres (16 in) or more of organic soil material in the upper 80 centimetres (31 in). Organic soil material has an organic carbon content (by weight) of 12 to 18 percent, or more, depending on the clay content of the soil. These materials include muck (sapric soil material), mucky peat (hemic soil material), or peat (fibric soil material). Aquic conditions or artificial drainage are required. Typically, histosols have very low bulk density and are poorly drained because the organic matter holds water very well. Most are acidic and many are very deficient in major plant nutrients which are washed away in the consistently moist soil.

Histosols are known by various other names in other countries, such as peat or muck. In the Australian Soil Classification, histosols are called Organosols.

Histosols form whenever organic matter forms at a more rapid rate than it is destroyed. This occurs because of restricted drainage precluding aerobic decomposition, and the remains of plants and animals remain within the soil. Thus, histosols are very important ecologically because they, and gelisols, store large quantities of organic carbon. If accumulation continues for a long enough period, coal forms.

Most histosols occur in Canada, Scandinavia, the West Siberian Plain, Sumatra, Borneo and New Guinea. Smaller areas are found in other parts of Europe, the Russian Far East (chiefly in Khabarovsk Krai and Amur Oblast), Florida and other areas of permanent swampland. Fossil histosols are known from the earliest extensive land vegetation in the Devonian.

Histosols are generally very difficult to cultivate because of the poor drainage and often low chemical fertility. However, histosols formed on very recent glacial lands can often be very productive when drained and produce high-grade pasture for dairying or beef cattle. They can sometimes be used for fruit if carefully managed, but there is a great risk of the organic matter becoming dry powder and eroding under the influence of drying winds. A tendency towards shrinkage and compaction is also evident with crops.

Like gelisols, histosols have greatly restricted use for civil engineering purposes because heavy structures tend to subside in the wet soil.

In USA soil taxonomy, histosols are subdivided into:

- **Folists** - histosols that are not saturated with water for long periods of time during the year.
- **Fibrists** - histosols that are primarily made up of only slightly decomposed organic materials, often called peat.
- **Hemists** - histosols that are primarily made up of moderately decomposed organic materials.
- **Saprists** - histosols that are primarily made up of highly decomposed organic materials, often called muck.

Inceptisol

Inceptisols are a soil order in USDA soil taxonomy. They form quickly through alteration of parent material. They are more developed than entisols. They have no accumulation of clays, iron oxide, aluminium oxide or organic matter. They have an ochric or umbric horizon and a cambic subsurface horizon.

Mollisol

Mollisols are a soil order in USDA soil taxonomy. Mollisols form in semi-arid to semi-humid areas, typically under a grassland cover. They are most commonly found in the mid-latitudes, namely in North America, mostly east of the Rocky Mountains, in South America in Argentina (Pampas) and Brazil, and in Asia in Mongolia and the RussianSteppes. Their parent material is typically base-rich and calcareous and include limestone, loess, or wind-blown sand. The main processes that lead to the formation of grassland Mollisols are melanisation, decomposition, humification and pedoturbation.

Mollisols have deep, high organic matter, nutrient-enriched surface soil (A horizon), typically between 60–80 cm in depth. This fertile surface horizon, known as a mollic epipedon, is the defining diagnostic feature of Mollisols. Mollic epipedons result from the long-term addition of organic materials derived from plant roots, and typically have soft, granular, soil structure.

Mollisols occur in savannahs and mountain valleys (such as Central Asia, or the North American Great Plains). These environments have historically been strongly influenced by fire and abundant pedoturbation from organisms such as ants and earth worms. It was estimated that in 2003, only 14 to 26 percent of grassland ecosystems still remained in a relatively natural state (that is, they were not used for agriculture due to the fertility of the A horizon). Globally, they represent ~7% of ice-free land area. As the world's most agriculturally productive soil order, the Mollisols represent one of the more economically important soil orders.

A Mollisol profile

Though most of the other soil orders known today existed by the time of the Carboniferous Ice Age 280 million years ago, Mollisols are not known from the paleopedological record any earlier than the *Eocene*. Their development is very closely associated with the cooling and drying of the global climate that occurred during the Oligocene, Miocene and Pliocene.

Suborders

Albolls — wet soils; aquic soil moisture regime with an eluvial horizon

Aquolls — wet soils; aquic soil moisture regime

Cryolls — cold climate; frigid or cryic soil temperature regime

Gelolls — very cold climate; mean annual soil temperature $<0°C$

Rendolls — lime parent material

Udolls — humid climate; udic moisture regime

Ustolls — subhumid climate; ustic moisture regime

Xerolls — Mediteranaean climate; xeric moisture regime

Soils which are in most ways similar to Mollisols but contain either continuous or discontinuous permafrost and are consequently affected by cryoturbation are common in the high mountain plateaux of Tibet and the Andean altiplano. Such soils are known as **Molliturbels** or **Mollorthels** and provide the best grazing land in such cold climates because they are not acidic like many other soils of very cold climates.

Other soils which have a mollic epipedon are classified as Vertisols because the presence of high shrink swell characteristics and relatively high clay contents takes precedence over the mollic epipedon. These are especially common in parts of South America in the Paraná River basin that have abundant but erratic rainfall and extensive deposition of clay-rich minerals from the Andes. Mollic epipedons also occur in some Andisols but the andic properties take precedence.

Suborders

- Anthrepts - modified by human habitation and farming
- Aquepts - with a water table close to the surface
- Cryepts - in cold climates
- Udepts - in humid climates
- Ustepts - in semiarid and sub-humid climates
- Xerepts - in areas with very dry summers and moist winters

Oxisol

Oxisols are an order in USDA soil taxonomy, best known for their occurrence in tropical rain forest, 15-25 degrees north and south of the Equator. They are classified as ferralsols in the World Reference Base for Soil Resources; some oxisols have been previously classified as laterite soils.

Oxisol profile

Formation

The main processes of soil formation of oxisols are weathering, humification and pedoturbation due to animals. These processes produce the characteristic soil profile. They are defined as soils containing *at all depths* no more than 10 percent weatherable minerals, and low cation exchange capacity. Oxisols are always a red or yellowish color, due to the high concentration of iron(III) and aluminium oxides and hydroxides. In addition they also contain quartz and kaolin, plus small amounts of other clay minerals and organic matter.

Etymology

The word "oxisol" comes from *"oxide"* in reference to the dominance of oxide minerals such as bauxite. In the FAO soil classification, oxisols are known as *ferralsols*.

Scientists originally thought that the heavy vegetation of tropical rain forests would provide rich nutrients, but as rainfall passes through the litter on the forest floor the rain is acidified and leaches minerals from the above soil layers. This forces plants to get their nutrition from decaying litter as oxisols are quite infertile due to the lack of organic matter and the almost complete absence of soluble minerals leached by the wet and humid climate.

Where it is Found

Present-day oxisols are found almost exclusively in tropical areas, in South America and Africa, almost always on highly stable continental cratons.

In Southeast Asia, oxisols are found on remnants of the Cimmerian microcontinent, and on the Shan–Thai Terrane. In Thailand, rhodic ferralsols, called Yasothon soils, are said to have formed under humid tropical conditions in the early Tertiary, on an extensive plain later uplifted to form the Khorat Plateau. Characterized by a bright red color, these relict soils occur on uplands in a great semicircle around the southern rim, overlying associated gravel horizons said to have been cleared of sand by termites, in a prolonged and still on-going process of bioturbation. Xanthic ferralsols of the Khorat and Udon Series, characterized by a pale yellow to brown color, developed in midlands in processes still under investigation; as are those forming lowland soils resembling European brown soils.

In Australia vast areas formerly covered in rainforest have become so dry that oxisols have formed a hard ironstone cover upon which only skeletal soils can form.

Genesis

Fossil oxisols are known from the first appearance of free oxygen in the atmosphere about 2,200 million years ago. In warm periods like the Mesozoic and Paleocene, oxisols extended to areas that now have quite cool climates, extending well into North America and Europe. It is believed oxisols became vegetated later than ultisols or alfisols, probably because vegetation took a long time to adapt to the infertility of oxisols.

Uses

Oxisols are often used for tropical crops such as cocoa and rubber. In some cases, rice is grown on them. Permanent cropping of oxisols in low-income areas is very difficult because of low cation exchange capacities and high phosphorus fixation on iron and aluminium oxides (ligand exchange mechanism; inner sphere complex with phosphate). However, many oxisols can be cultivated over a wide range of moisture conditions. On this account, oxisols are intensively exploited for agriculture in some regions which have enough wealth to support modern agricultural practices (including regular additions of lime and fertilizer). A recent example of exploitation by modern methods involves the growing of soybeans in Brazil.

Suborders

Oxisols are divided into the following suborders:

- **Aquox** - oxisols with a water table at or near the surface for much of the year
- **Perox** - oxisols of continuously humid climates, where precipitation exceeds evapotranspiration in all months
- **Torrox** - oxisols of arid climates. Because the present climate can never produce enough weathering to produce oxisols, torrox soils are always paleosols formed during periods of much wetter climates. They occur mainly in Southern Africa.

- **Ustox** - oxisols of semiarid and subhumid climates
- **Udox** - oxisols of humid climates

Podzol

In soil science, **podzols** (known as **spodosols** in China and the United States of America and **podosols** in Australia) are the typical soils of coniferous, or boreal forests. They are also the typical soils of eucalypt forests and heathlands in southern Australia, while in Western Europe podzols develop on heathland, which is often a construct of human interference through grazing and burning. Many podzols in this region may have developed over the past 3000 years in response to vegetation and climatic changes. In some British moorlands with podzolic soils there are brown earths preserved under Bronze Age barrows (Dimbleby, 1962). Podzol means under-ash, and is derived from the Russian под (pod) + зола́ (zola); the full form is подзо́листая по́чва (podzolistaya pochva, under-ashed soil). It likely refers to the common experience of Russian peasants of plowing up an apparent under-layer of ash (leached or E horizon) during first plowing of a virgin soil of this type.

Podzols are able to occur on almost any parent material but generally derive from either quartz-rich sands and sandstones or sedimentary debris from magmatic rocks, provided there is high precipitation. Most podzols are poor soils for agriculture due to the sandy portion, resulting in a low level of moisture and nutrients. Some are sandy and excessively drained. Others have shallow rooting zones and poor drainage due to subsoil cementation. A low pH further compounds issues, along with phosphate deficiencies and aluminium toxicity. The best agricultural use of podzols is for grazing, although well-drained loamy types can be very productive for crops if lime and fertilizer are used.

The E horizon, which is usually 4 to 8 centimetres (1.57 to 3.15 in) thick, is low in Fe and Al oxides and humus. It is formed under moist, cool and acidic conditions, especially where the parent material, such as granite or sandstone, is rich in quartz. It is found under a layer of organic material in the process of decomposition, which is usually 5 to 10 centimetres (1.97 to 3.94 in) thick. In the middle, there is often a thin layer of 0.5 to 1 centimetre (0.2 to 0.4 in). The bleached soil goes over into a red or redbrown horizon called rusty soil. The colour is strongest in the upper part, and change at a depth of 50 to 100 centimetres (19.7 to 39.4 in) progressively to the part of the soil that is mainly not affected by processes; that is the parent material. The soil profiles are designated by the letters A (topsoil), E (eluviated soil), B (subsoil) and C (parent material).

In some podzols, the E horizon is absent—either masked by biological activity or obliterated by disturbance. Podzols with little or no E horizon development are often classified as *brown podzolic* soils, also called **umbrisols** or **umbrepts**.

The picture is of a stagnopodzol in upland Wales, and shows the typical sequence of organic topsoil with leached grey-white subsoil with iron-rich horizon below. The example has two weak ironpans.

Geographic Distribution

Podzols cover about 4,850,000 square kilometres (1,870,000 sq mi) worldwide and are usually found under sclerophyllous woody vegetation. By extent podzols are most common in temperate and boreal zones of the northern hemisphere but they can also be found in other settings including both temperate rainforests and tropical areas.

In South America podzols occur beneath *Nothofagus betuloides* forests in Tierra del Fuego.

Podzolisation

The main process in the formation of podzols is podzolisation. Podzolisation is a complex process (or number of sub-processes) in which organic material and soluble minerals (commonly iron and aluminium) are leached from the A and E horizons to the B horizon. It involves mobilization and precipitation of dissolved organic matter, together with aluminum and iron as they leach down from the A and E horizons to the B horizon. Through this process the overlying eluvial horizons are bleached. The complexes move to the brown, red or black horizon, which consist of cemented sesquioxides and/or organic compounds. Podzolisation occurs under low pH values.

In podzols, translocation has meant the eluviation of clays, humic acids, iron, and other soluble constituents from the A and E horizons. These constituents may then accumulate to form a spodic illuvial horizon and in some cases a placic horizon or iron band. Podzolization occurs when severe leaching leaves the upper horizon virtually depleted of all soil constituents except quartz grains. Clay minerals in the A horizon decompose by reaction with humic acids and form soluble salts. The leached material from the A horizon is deposited in the B horizon as a humus-rich horizon band, a hard layer of sesquioxides or a combination of the two.

These sub-processes include mobilisation, eluviation and illuviation. Mobilisation and eluviation both move organic materials and minerals through the A horizon into the B horizon. During this, they react with the water (illuviation) to become oxidised. This process of podzolisation results in the characteristic soil profile of spodosols, in which the E horizon is usually an ashen grey or white colour without structure and there is a distinctive hardpan oxide layer in the B horizon (which is always darker than the E horizon). The E horizon can be dark grey in profiles which are high in organic matter, but in such cases the underlying B is *very* dark. Spodosols over mafic rock may have a medium brown E horizon underlain by a darker B.

However, as conifers allelopathically reduce competition by producing a thick O horizon of mor (acidic and poisonous leaf litter that is slow to decompose), the primary form of plant-soil interactions is that of the conifers themselves. The acidic O horizon, along with rainfall patterns that are similar to that of the moister grasslands, also promotes the illuviation of oxides of aluminium and iron.

Spodosols are rare as paleosols. Though they are known from as far back as the Carboniferous, there are few examples surviving from before the first Pleistocene glaciation, and some of these may not be true spodosols.

Ultisol

Ultisols, commonly known as **red clay soils**, are one of twelve soil orders in the United States Department of Agriculture soil taxonomy. They are defined as mineral soils which contain no calcareous material anywhere within the soil, have less than 10% weatherable minerals in the extreme top layer of soil, and have less than 35% base saturation throughout the soil. Ultisols occur in humid temperate or tropical regions. While the term is usually applied to the red clay soils of the Southern United States, ultisols are also found in regions of Africa, Asia, and South America. In the World Reference Base for Soil Resources system, most ultisols are known as acrisols. Others with higher-activity clays are classed as alisols or nitisols.

The word ultisol is derived from ultimate, because ultisols were seen as the ultimate product of continuous weathering of minerals in a humid, temperate climate without new soil formation via glaciation.

Ultisols vary in color from purplish-red, to a bright reddish-orange, to pale yellowish-orange and even some subdued yellowish-brown tones. They are typically quite acidic, often having a pH of less than 5. The red and yellow colors result from the accumulation of iron oxide (rust), which is highly insoluble in water. Major nutrients, such as calcium and potassium, are typically deficient in ultisols, which means they generally cannot be used for sedentary agriculture without the aid of lime and other fertilizers, such as superphosphate. They can be easily exhausted, and require more careful management than alfisols or mollisols. However, they can be cultivated over a relatively wide range of moisture conditions.

Ultisols can have a variety of clay minerals, but in many cases the dominant mineral is kaolinite. This clay has good bearing capacity and no shrink–swell property. Consequently, well-drained kaolinitic ultisols such as the Cecil series are suitable for urban development.

Ultisols are the dominant soils in the Southern United States (where the Cecil series is most famous), southeastern China, Southeast Asia, and some other subtropical and tropical areas. Their northern limit (except fossil soils) is very sharply defined in North America by the limits of maximum glaciation during the Pleistocene, because ultisols typically take hundreds of thousands of years to form—far longer than the length of an interglacial period today.

The oldest fossil ultisols are known from the Carboniferous period when forests first developed. Though known from far north of their present range as recently as the Miocene, ultisols are surprisingly rare as fossils overall, since they would have been expected to be very common in the warm Mesozoic and Tertiary paleoclimates.

An Ultisol profile.

Red clay soil is common throughout the Southern United States, especially around the Piedmont. This photo was taken in North Carolina.

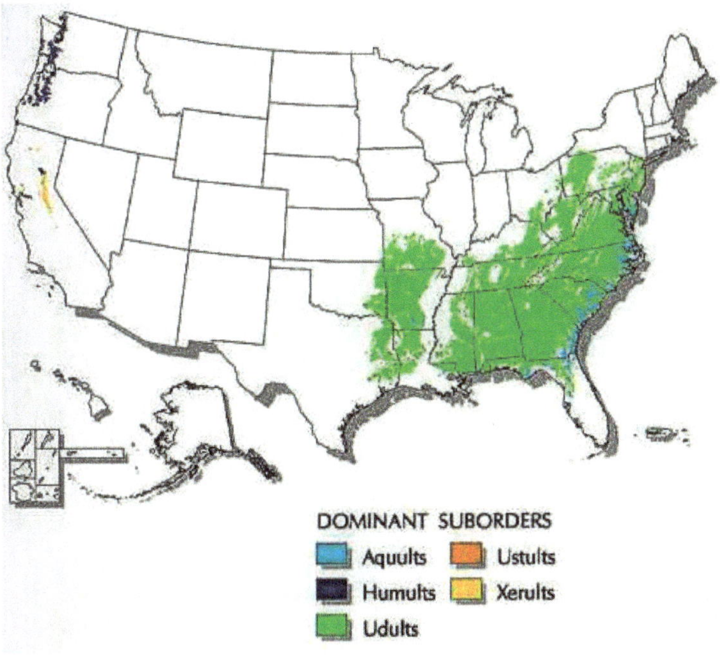

Map showing distribution and types of ultisols throughout the United States; there is no ultisol on the Ohio River flood plains, as the river has historically deposited other soil types there during its regular natural flooding.

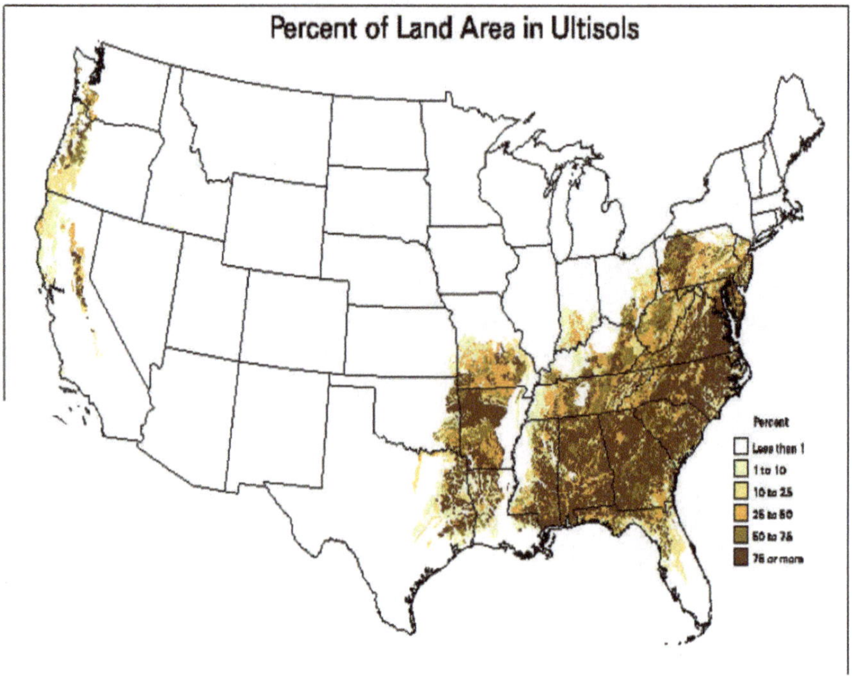

Map of the United States showing what percentage of the soil in a given area is classified as an ultisol-type soil. The great majority of the land area classified in the highest category (75%-or-greater ultisol) lies in the South and overlays with the Piedmont Plateau, which runs as a diagonal line through the South from southeast (in Alabama) to northwest (up through Virginia, and even into parts of Maryland).

Suborders

- **Aquults**: Ultisols with a water table at or near the surface for much of the year
- **Humults**: well-drained ultisols that have high organic matter content
- **Udults**: Ultisols of humid climates
- **Ustults**: Ultisols of semiarid and subhumid climates
- **Xerults**: temperate ultisols with very dry summers and moist winters

Vertisol

In both the FAO and USDA soil taxonomy, a **vertisol** (Vertosol in the Australian Soil Classification) is a soil in which there is a high content of expansive clay known as montmorillonite that forms deep cracks in drier seasons or years. Alternate shrinking and swelling causes *self-mulching*, where the soil material consistently mixes itself, causing vertisols to have an extremely deep A horizon and no B horizon. (A soil with no B horizon is called an *A/C soil*). This heaving of the underlying material to the surface often creates a microrelief known as *gilgai*.

Vertisols typically form from highly basic rocks, such as basalt, in climates that are seasonally humid or subject to erratic droughts and floods, or that impeded drainage. Depending on the parent material and the climate, they can range from grey or red to the more familiar deep black (known as "black earths" in Australia, "black gumbo" in East Texas, and "black cotton" soils in East Africa).

Vertisols are found between 50°N and 45°S of the equator. Major areas where vertisols are dominant are eastern Australia (especially inland Queensland and New South Wales), the Deccan Plateau of India, and parts of southern Sudan, Ethiopia, Kenya, and Chad (the *Gezira*), and the lower Paraná River in South America. Other areas where vertisols are dominant include southern Texas and adjacent Mexico, central India, northeast Nigeria, Thrace, New Caledonia and parts of eastern China.

The natural vegetation of vertisols is grassland, savanna, or grassy woodland. The heavy texture and unstable behaviour of the soil makes it difficult for many tree species to grow, and forest is uncommon.

The shrinking and swelling of vertisols can damage buildings and roads, leading to extensive subsidence. Vertisols are generally used for grazing of cattle or sheep. It is not unknown for livestock to be injured through falling into cracks in dry periods. Conversely, many wild and domestic ungulates do not like to move on this soil when inundated. However, the shrink-swell activity allows rapid recovery from compaction.

When irrigation is available, crops such as cotton, wheat, sorghum and rice can be grown. Vertisols are especially suitable for rice because they are almost impermeable when saturated. Rainfed farming is very difficult because vertisols can be worked only under a very narrow range of moisture conditions: they are very hard when dry and very sticky when wet. However, in Australia, vertisols are highly regarded, because they are among the few soils that are not acutely deficient in available phosphorus. Some, known as "crusty vertisols", have a thin, hard crust when dry that can persist for two to three years before they have crumbled enough to permit seeding.

In the USA soil taxonomy, vertisols are subdivided into:

- **Aquerts**: Vertisols which are subdued aquic conditions for some time in most years and show redoximorphic features are grouped as Aquerts. Because of the high clay content, the permeability is slowed down and aquic conditions are likely to occur. In general, when precipitation exceeds evapotranspiration, ponding may occur. Under wet soil moisture conditions, iron and manganese are mobilized and reduced. The manganese may be partly responsible for the dark color of the soil profile.
- **Cryerts** (not classified as vertisols in the FAO classification): They have a cryic soil temperature regime. Cryerts are most extensive in the grassland and forest-grassland transitions zones of the Canadian Prairies and at similar latitudes in Russia.
- **Xererts**: They have a thermic, mesic, or frigid soil temperature regime. They show cracks that are open at least 60 consecutive days during the summer, but are closed at least 60 consecutive days during winter. Xererts are most extensive in the eastern Mediterranean and parts of California.

- **Torrerts**: They have cracks that are closed for less than 60 consecutive days when the soil temperature at 50 cm is above 8 °C. These soils are not extensive in the U.S., and occur mostly in west Texas, New Mexico, Arizona, and South Dakota, but are the most extensive suborder of Vertisols in Australia.
- **Usterts**: They have cracks that are open for at least 90 cumulative days per year. Globally, this suborder is the most extensive of the Vertisols order, encompassing the Vertisols of the tropics and monsoonal climates in Australia, India, and Africa. In the U.S. the Usterts are common in Texas, Montana, Hawaii, and California.
- **Uderts**: They have cracks that are open less than 90 cumulative days per year and less than 60 consecutive days during the summer. In some areas, cracks open only in drought years. Uderts are of small extent globally, being most abundant in Uruguay and eastern Argentina, but also found in parts of Queensland and the "Black Belt" of Mississippi and Alabama.

A Vertisol profile.

WORLD REFERENCE BASE FOR SOIL RESOURCES

The World Reference Base for Soil Resources (WRB) is the international standard taxonomic soil classification system endorsed by the International Union of Soil Sciences (IUSS). It was developed by an international collaboration coordinated by the International Soil Reference and Information Centre (ISRIC) and sponsored by the IUSS and the FAO via its Land & Water Development division. It replaces the previous FAO soil classification.

The WRB borrows heavily from modern soil classification concepts, including USDA soil taxonomy, the legend for the FAO Soil Map of the World 1988, the Référentiel Pédologique and Russian concepts. The classification is based mainly on soil morphology as an expression of pedogenesis. A major difference with USDA soil taxonomy is that soil climate is not part of the system, except insofar as climate influences soil profile characteristics. As far as possible, diagnostic criteria match those of existing systems, so that correlation with national and previous international systems is as straightforward as possible.

The WRB is meant for correlation of national and local systems. The level of detail corresponds to USDA soil taxonomy subgroups, without the soil climate information. The second edition was not detailed enough for mapping at scales larger than about 1:200 000, and a third edition has been published, improving the system for soil mapping.

Soil Groups

Following is a highly simplified description of each reference soil group.

Code	Soil type	Brief description
AC	Acrisols	Red, brown or yellow coloured soil, develops in areas of intense weathering, has a clay rich B horizon
AB	Albeluvisols	Obsolete, replaced in 3rd edition by Retisols (c.f.)
AL	Alisols	
AN	Andosols	Soil developed from volcanic material, are young immature soils, characteristics depend on type of volcanic material
AT	Anthrosols	Soils that have been modified profoundly through human activities, such as addition of organic or mineral material, charcoal or household wastes, or irrigation and cultivation.
AR	Arenosols	Sandy soil with no more profile development than an A horizon
CL	Calcisols	Soil with a substantial secondary accumulation of lime
CM	Cambisols	Transformation of soil matter (Fe particularly) in situ without moving in profile. Mostly brownish color.
CH	Chernozem	Fertile black-coloured soil containing a high percentage of humus, phosphoric acids, phosphorus and ammonia
CR	Cryosols	Soil in permafrost areas, exhibits cryoturbation and is usually rich in organic matter
DU	Durisol	Soil of some arid and semi-arid environments, contains cemented secondary silica
FR	Ferralsols	Red to yellow soil rich in iron and aluminium, common in temperate to tropical humid areas
FL	Fluvisol	Soil developed above flood plain sediments, A horizon is commonly directly above C horizon
GL	Gleysols	
GY	Gypsisols	
HS	Histosols	Soil consisting primarily of organic materials, common in wetlands
KS	Kastanozem	
LP	Leptosols	Shallow soil over bedrock, calcareous material or a deeper soil that is gravelly or stony, common in mountains

LX	Lixisols	
LV	Luvisols	
NT	Nitisols	
PH	Phaeozem	Sod organic-accumulative
PL	Planosols	
PT	Plinthosols	
PZ	Podzols	Soil that presents significant podzolization, common in coniferous forests
RG	Regosols	
RT	Retisols	Soils having a clay illuviation horizon (similar to Luvisols), but with an interfingering of bleached coarser- textured soil material into the illuviation horizon forming a net-like pattern.
SC	Solonchaks	
SN	Solonetz	
TC	Techosols	soils whose properties and pedogenesis are dominated by their technical origin.
UM	Umbrisols	Soil with a dark topsoil and in which organic matter has accumulated significantly within the mineral surface soil
VR	Vertisols	Shows significant and recurrent swelling with water, high content of expansive clay

Acrisol

An **acrisol** is a type of soil as classified by the Food and Agriculture Organization. It is clay-rich, and is associated with humid, tropical climates, such as those found in Brazil, and often supports forested areas. It is one of the 30 major soil groups of the World Reference Base for Soil Resources. Acrisols correspond to the aquult, humults, udults and ustults sub-orders of ultisols in the USDA soil taxonomy and also to oxisols with a candic horizons and some alfisols.The acrisols low fertility and toxic amounts of aluminium pose limitations to its agricultural use, favouring in many places its use for silviculture, low intensity pasture and protected areas. Crops that can be successfully cultivated, if climate allows, include tea, rubber tree, oil palm, coffee and sugar cane.

The types of acrisol are:

- ACf Ferric Acrisols
- ACg Gleyic Acrisols
- ACh Haplic Acrisols
- ACp Plinthic Acrisols
- ACu Humic Acrisols

A gleyic acrisol profile from Malaysia.

Albeluvisols

An **albeluvisol** in the FAO World Reference Base for Soil Resources is a soil with a thin, dark surface horizon on a bleached subsurface horizon (an albic horizon) that tongues into a clay illuviation (Bt) horizon. The Bt horizon has an irregular or broken upper boundary resulting from the tonguing of bleached soil material into the illuviation horizon. Albeluvisols correlate with Glossaqualfs, Glossocryalfs and Glossudalfs in the USDA soil taxonomy.

These soils are formed mostly in unconsolidated glacial till, lacustrine or fluvial materials or aeolian deposits such as loess. They occur on flat to undulating plains under coniferous forest or mixed forest in boreal and temperate climates with cold winters and short cool summers.

The agricultural suitability of Albeluvisols is limited because of their acidity, low nutrient levels, tillage and drainage problems. In northern regions there is also a short growing season and severe frost during the long winter. The Albeluvisols of the northern taiga zone are almost exclusively under forest with small areas used for pasture or hay fields. In the southern taiga zone, less than 10 percent of the non-forested area is used for livestock farming. In the southern and western parts of the taiga in Russia arable crops, such as cereals, potatoes, sugar beet and forage maize, are found, especially on soils with higher base saturations in the subsoil.

Albeluvisols cover an estimated 320 million ha in Europe, North Asia, Central Asia and in North America. They are concentrated in two regions:

- the continental regions that had permafrost in the Pleistocene of northeast Europe, northwest Asia and southern Canada, which constitute by far the largest areas of Albeluvisols;
- the loess and cover sand areas and old alluvial areas in moist temperate regions, such as France, central Belgium, the southeast of the Netherlands and the north and north-east of Germany.

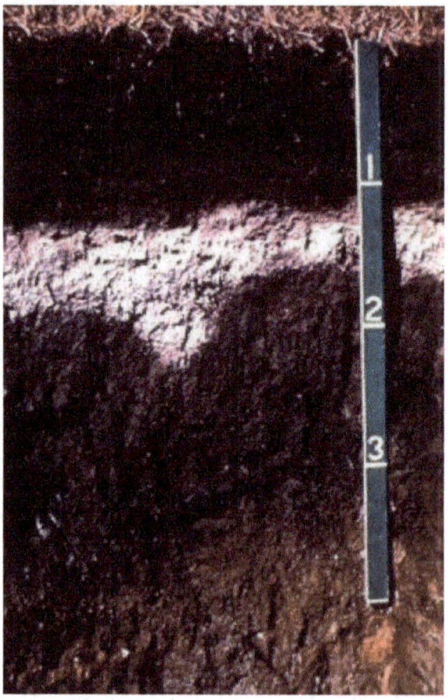

An Albic horizon.

Alisols

Alisols define a soil group within the World Reference Base for Soil Resources

Properties include having an argic horizon, which has a specific cation exchange capacity; a predominantly alic properties zone between 250mm and 1000 mm from the soil surface; and no diagnostic horizons other than an ochric, umbric, albic, andic, ferric, nitic, plinthic or vertic horizon.

Simply put, alisols are poorly drained soils with a dense subsurface clay layer, which causes a relatively high concentration of aluminum ions in the root zone. There exist mixed forms, for example ‹gleyic alisol›, that are mainly alisol, but also contain components that are found in gleysols.

Alisols occur mainly in tropical and humid subtropical climates, though a few are found in unglaciated landscapes as far north as Brittany. Compared to Acrisols and

Ferralsols, Alisols have much higher-activity clays and are likely to be found on younger terrains or more geologically active regions such as Kyushu and Chugoku.

Agricultural Use

Alisols are acidic (increased by limited drainage) and therefore need liming, contain few nutrients and therefore need fertilizer, and do not have much surface coherence so are easily eroded. Aluminium and manganese toxicity is a very serious problem in Alisols, because at the low pH of these soils such generally insoluble metals become soluble and can poison plants which are not tolerant of them. Encyclopædia Britannica mentions oil palm, cotton, and maize (corn) as crops suitable to be grown on Alisols, though most crops require very intensive fertilisation for long-term success.

Andisol

In USDA soil taxonomy, andisols are soils formed in volcanic ash and defined as soils containing high proportions of glass and amorphous colloidal materials, including allophane, imogolite and ferrihydrite. In the FAO soil classification, andisols are known as andosols.

Because they are generally quite young, andisols typically are very fertile except in cases where phosphorus is easily fixed (this sometimes occurs in the tropics). They can usually support intensive cropping, with areas used for wet rice in Java supporting some of the densest populations in the world. Other andisol areas support crops of fruit, maize, tea, coffee or tobacco. In the Pacific Northwest USA, andisols support very productive forests.

Andisols occupy about 1% of the global ice-free land area. Most occur around the Pacific Ring of Fire, with the largest areas found in central Chile, Ecuador, Colombia, Mexico, the Pacific Northwest USA, Japan, Java and New Zealand›s North Island. Other areas occur in the Great Rift Valley, Kenya, Italy, Iceland and Hawai.

Fossil andisols are known from areas far from present-day volcanic activity and have in some cases been dated as far back as the Precambrian 1.5 billion years ago.

Suborders

- **Aquands** - Andisols with a water table at or near the surface for much of the year.
- **Gelands** - Andisols of very cold climates (mean annual temperature <0°C).
- **Cryands** - Andisols of cold climates.
- **Torrands** - Andisols of very dry climates.
- **Ustands** - Andisols of semiarid and sub humid climates.
- **Udands** - Andisols of humid climates.
- **Xerands** - temperate Andisols with very dry summers and moist winters.
- **Vitrands** - relatively young Andisols that are coarse-textured and dominated by glass.

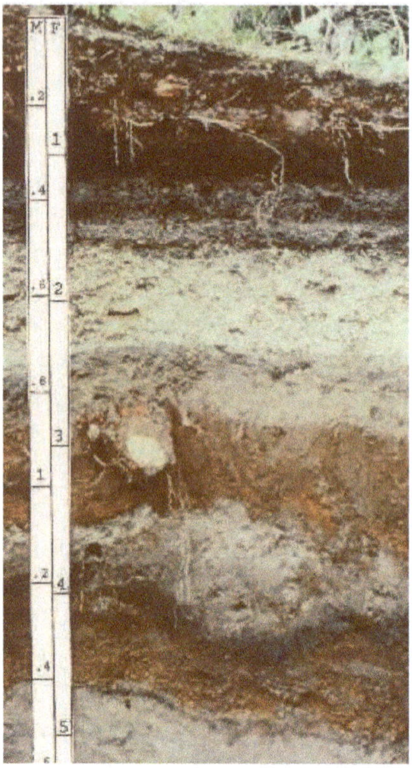

Anthrosol

An Anthrosol in the FAO World Reference Base for Soil Resources is a type of soil that has been formed or heavily modified due to long-term human activity, such as from irrigation, addition of organic waste or wet-field cultivation used to create paddy fields.

Such soils can be formed from any parent soil, and are commonly found in areas where agriculture has been practised for centuries. Anthrosols can be found worldwide, though they tend to have different soil horizons in different regions. For example, in northwestern Europe Anthrosols commonly have plaggic or terric horizons, and together they cover some 500,000 hectares.

Anthrosols in Archaeology

The presence of anthrosols can be used to detect long-term human habitation, and has been used by archaeologists to identify sites of interest. Anthrosols that can indicate such activity can be described as, for instance, plaggic (from the long-terms use of manure to enrich soil), irragic (from the use of flood or surface irrigation), hortic (from deep cultivation, manure use and presence of other anthropogenic organic matter such as kitchen waste), anthraquic (from anthros - man - and aqua - water - meaning produced by man-made soil moisture management including irrigation or terracing), etc. Anthrosols can be detected by visual inspection of soils, or even from satellite imagery.

Psamment

In USDA soil taxonomy, a **psamment** is defined as an entisol which consists basically of unconsolidated sand deposits, often found in shifting sand dunes but also in areas of very coarse-textured parent material subject to millions of years of weathering. This latter case is characteristic of the Guiana Highlands of northern South America. A psamment has no distinct soil horizons, and must consist entirely of material of loamy sand or coarser in texture. In the FAO soil classification, psamments are known as **arenosols**.

Psamments cover 3.4% of the global land mass. They occur throughout the world, being especially abundant in the deserts of Africa and Australia and on the ancient landforms of eastern South America. Areas dominated by psamments also occur in other humid regions, notably in Florida and Nebraska (the Sand Hills).

Psamments typically have very low water-holding capacities because the sand in the soil is not *graded* so that sands of varying coarseness are constantly mixed right through the soil. Because most sands are highly siliceous, psamments are also extremely low in all essential nutrients, most especially phosphorus and are highly acidic in all except very arid climates. Psamments formed as a result of glacial erosion (common in northern Europe) are typically of rather higher native fertility because of their youth, but are still much less fertile than most soils in the regions they are located in.

The vegetation on psamments varies enormously owing to the variety of climates, but in many cases is remarkably well-adapted to the climate, as with the *Fynbos* of the Cape region in South Africa - famous for its remarkable biodiversity and the equally species-rich Kwongan of southwestern Western Australia. The *campinas* and *kerangas* are healthy forests typical of psamments in South America and Borneo.

Few psamments are farmed, and where they are, the cost is high because of the expense of fertilization. They are always much less productive than other soils in the same region even when fertilized, and require careful management because the sand is very easily eroded.

In USDA soil taxonomy, psamments are divided into:

- *Cryopsamments:* psamments that have a cryic soil temperature regime
- *Quartzipsamments:* other psamments that have, in the 0.02 to 2.0 mm fraction within the particle-size control section, a total of more than 90 percent (by weighted average) resistant minerals
- *Torripsamments:* other psamments that have an aridic (or torric) moisture regime
- *Ustipsamments:* other psamments that have an ustic moisture regime
- *Xeropsamments:* other psamments that have a xeric moisture regime
- *Udipsamments:* other psamments.

Calcisol

A **Calcisol** in the FAO World Reference Base for Soil Resources is a soil with a substantial secondary accumulation of lime. Calcisols are common in calcareous parent materials and widespread in arid and semi-arid environments. Formerly Calcisols were internationally known as *Desert soils* and *Takyrs*. Calcisols are developed in mostly alluvial, colluvial and aeolian deposits of base-rich weathering material. They are found on level to hilly land in arid and semi-arid regions. The natural vegetation is sparse and dominated by xerophytic shrubs and trees and/or ephemeral grasses.

Dryness, and in places also stoniness and/or the presence of a shallow petrocalcic horizon, limit the suitability of Calcisols for agriculture. If irrigated, drained (to prevent salinisation) and fertilised, Calcisols can be highly productive under a wide variety of crops. Hilly areas with Calcisols are predominantly used for low volume grazing of cattle, sheep and goats.

Many Calcisols occur together with Solonchaks that are actually salt-affected Calcisols and/or with other soils with secondary accumulation of lime that do not key out as Calcisols. The total Calcisol area may well amount to some 10 million square kilometres, nearly all of it in the arid and semi-arid (sub)tropics of both hemispheres.

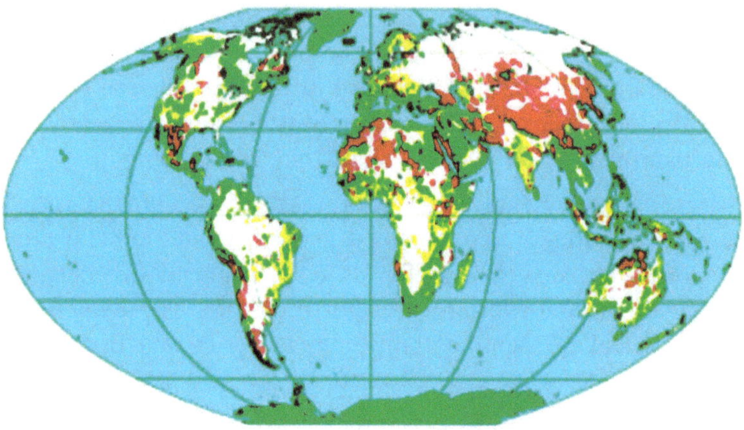

Distribution of Calcisols.

Cambisol

A **Cambisol** in the FAO World Reference Base for Soil Resources is a soil with a beginning of soil formation. The horizon differentiation is weak. This is evident from weak, mostly brownish discolouration and/or structure formation in the soil profile.

Cambisols are developed in medium and fine-textured materials derived from a wide range of rocks, mostly in alluvial, colluvial and aeolian deposits.

Most of these soils make good agricultural land and are intensively used. Cambisols in temperate climates are among the most productive soils on earth.

Cambisols cover an estimated 15 million square kilometres worldwide. They are well represented in temperate and boreal regions that were under the influence of glaciation during the Pleistocene, partly because the soil›s parent material is still young, but also because soil formation is comparatively slow in the cool, northern regions. Cambisols are less common in the tropics and subtropics. But they are common in areas with active erosion where they may occur in association with mature tropical soils.

Chernozem

Chernozem is a black-coloured soil containing a high percentage of humus (7% to 15%), and high percentages of phosphoric acids, phosphorus and ammonia. Chernozem is very fertile and produces a high agricultural yield.

Chernozem field in Black Dirt Region of Orange County, New York, United States

Distribution

There are two "chernozem belts" in the world: the Eurasian steppe which extends from eastern Croatia (Slavonia), along the Danube (northern Serbia, northern Bulgaria (Danubian Plain), southern Romania (Wallachian Plain) and Moldova) to northeast Ukraine across the Black Earth Region and southern Russia into Siberia, and the other from the Canadian Prairies in Manitoba through the Great Plains of the United States as far south as Kansas. Similar soil types occur in Texas and Hungary. Chernozem layer thickness may vary widely, from several inches up to 60 inches (1.5 metres) in Ukraine.

The terrain can also be found in small quantities elsewhere (for example, on 1% of Poland). It also exists in Northeast China, near Harbin. The only true chernozem in Australia is located around Nimmitabel producing some of the richest soils in the nation.

The sale of agricultural land has been illegal in Ukraine since 1992, but there is a black market for chernozem soil, transported by the truckload, with approximately US$900 million in annual sales.

Gelisol

Gelisols are an order in USDA soil taxonomy. They are soils of very cold climates which are defined as containing permafrost within two meters of the soil surface. The word "gelisol comes from the Latin *gelare* meaning "to freeze", a reference to the process of cryoturbation that occurs from the alternating thawing and freezing characteristic of gelisols.

In United Nations Food and Agriculture Organization soil classification system, gelisols are known as **cryosols**.

Structurally, gelisols have no B horizon and have an A horizon resting on the permafrost. Because soil organic matter accumulates in the upper layer, most gelisols are black or dark brown in soil color, followed by a shallow mineral layer. Despite the influence of glaciation in most areas where gelisols occur, chemically they are not highly fertile because nutrients, especially calcium and potassium, are very easily leached above the permafrost. The permafrost greatly restricts the engineering use of gelisols, as large structures (e.g. buildings) subside as the frozen earth thaws when they are put in place.

Gelisols are found chiefly in Siberia, Alaska and Canada. Smaller areas are found in the Andes (mainly near the intersection between Chile, Bolivia and Argentina), Tibet, northern Scandinavia and the ice-free parts of Greenland and Antarctica. Fossil gelisols are known from as far back as Precambrian ice ages 900 million years ago.

Suborders

In USDA soil taxonomy, gelisols are subdivided into:

- **Histels**: organic soils similar to histosols except that they have permafrost within two meters below ground surface. They have 80% or more organic materials from the soil surface to a depth of 50 cm or to a glacic layer or densic, lithic, or paralithic contact, whichever is shallowest. These soils occur predominantely in subarctic and low arctic regions of continuous or widespread permafrost. Less than one-third of the active layer (the soil between the ground surface and a permafrost table) or an ice layer which is at least 30-cm thick has been cryoturbated.
- **Turbels**: soils that show marked influence of cryoturbation (more than one-third of the depth of the active layer) such as irregular, broken, or distorted horizon boundaries and involutions and areas with patterned ground. They commonly contain tongues of mineral and organic horizons, organic and mineral intrusions and oriented rock fragments. Organic matter is accumulated on top of the permafrost and ice wedges are a common features. Turbels occur primarily in the zone of continuous permafrost.
- **Orthels**: soils that show little or no cryoturbation (less than one-third of the depth of the active layer). Patterned ground (except for polygons) generally is lacking. Orthels occur primarily within the zone of discontinuous permafrost, and in alpine areas.

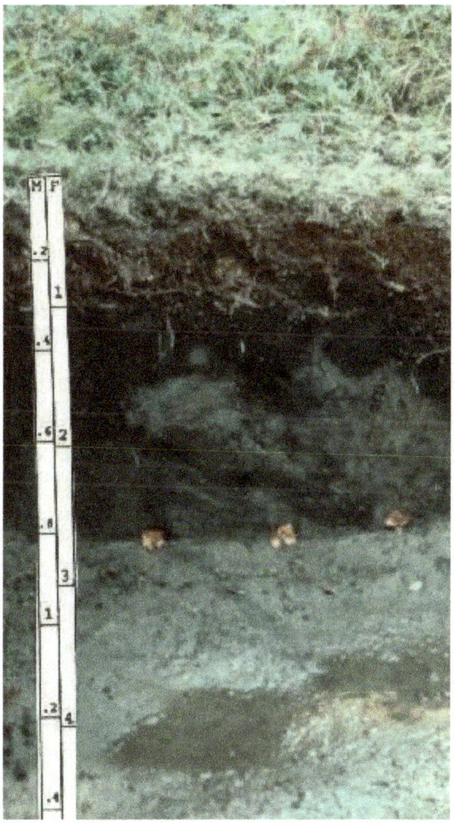

A Gelisol profile.

Durisol

In the FAO World Reference Base for Soil Resources, a **durisol** is a very shallow to moderately deep, free-draining soil of arid and semi-arid environments, that contains cemented secondary silica (SiO_2) in the upper metre of soil. Durisols are internationally known as "hardpan soils" (Australia) or "dorbank" (South Africa) or they represent the "duripan phase" of other soils, e.g. of Calcisols (FAO).

Durisols are developed mainly in alluvial and colluvial deposits of all texture classes. They are found on level and slightly sloping alluvial plains, terraces and gently sloping piedmont plains in arid, semi-arid and Mediterranean regions.

The soils have AC or ABC profile. Eroded Durisols with exposed hard horizons (a *petroduric horizon*) are common in (gently) sloping terrain.

Most Durisols can only be used for extensive grazing. Arable cropping of Durisols is limited to areas where irrigation water is available.

Extensive areas of Durisols occur in Australia, in South Africa, Namibia and in the United States (notably in Nevada, California and Arizona); minor occurrences have been reported from Central and South America and from Kuwait. Durisols are a new introduction in international soil classification and have not often been mapped as such. A precise indication of their extent is not (yet) available.

Fluvisol

A **Fluvisol** in the FAO World Reference Base for Soil Resources is a genetically young soil in alluvial deposits . Apart from river sediments, they also occur in lacustrine and marine deposits. Fluvisols correlate with Fluvents and Fluvaquents of the USDA soil taxonomy. The good natural fertility of most Fluvisols and their attractive dwelling sites on river levees and higher parts in marine landscapes were recognized in prehistoric times.

Fluvisols are found on alluvial plains, river fans, valleys and tidal marshes on all continents and in all climate zones. Under natural conditions periodical flooding is fairly common. The soils have a clear evidence of stratification. Soil horizons are weakly developed, but a distinct topsoil horizon may be present.

Many dryland crops are grown on Fluvisols, normally with some form of water control. On tropical Fluvisols with satisfactory irrigation and drainage paddy rice cultivation is widespread.

Some coastal lowlands have Fluvisols with sulphidic material within the profile (a thionic horizon). These Acid Sulphate Soils have severe constraints for agricultural use as low pH-values, toxic aluminium levels and high concentrations of salts. Tidal lands that are strongly saline are best kept under mangroves or some other salt-vegetation.

Major concentrations of Fluvisols are found along rivers and lakes, e.g. in the Amazon Basin, the Ganges Plain of India, the plains near Lake Chad in Central Africa, and the marshlands of Brazil, Paraguay and northern Argentina. Also in deltaic areas, e.g. the deltas of the Ganges–Brahmaputra, Indus, Mekong, Mississippi, Nile, Niger, Orinoco, Po, Rhine, Rio de la Plata, and Zambezi. Furthermore in areas of recent marine deposits, e.g. the coastal lowlands of Sumatra, Kalimantan, Irian Jaya and Papua New Guinea.

Acid Sulphate Soils are found in the coastal lowlands of Southeast Asia, West Africa and along the northeast coast of South America (French Guiana, Guyana, Suriname and Venezuela).

Gleysol

A Gley (Russian: gley is dialectical word, literally clay) is a wetland soil (hydric soil) that, unless drained, is saturated with groundwater for long enough periods to develop a characteristic gleyic colour pattern. This pattern is essentially made up of reddish, brownish or yellowish colours at surfaces of soil particles (*peds*) and/or in the upper soil horizons mixed with greyish/blueish colours inside the peds and/or deeper in the soil. Gleysols are also known as *Gleyzems* and *meadow soils* (Russia), *Aqu*-suborders of Entisols, Inceptisols and Mollisols (USDA soil taxonomy), or as *groundwater soils* and *hydro-morphic soils*.

Gleysols occur on wide range of unconsolidated materials, mainly fluvial, marine and lacustrine sediments of Pleistocene or Holocene age, with basic to acidic mineralogy. They are found in depression areas and low landscape positions with shallow groundwater.

Wetness is the main limitation of virgin Gleysols; these are covered with natural swamp vegetation and lie idle or are used for extensive grazing. Artificially drained Gleysols are used for arable cropping, dairy farming and horticulture. Gleysols in the tropics and subtropics are widely planted to rice.

Gleysols occupy an estimated 720 million hectares worldwide. They are azonal soils and occur in nearly all climates. The largest extent of Gleysols is in northern Russia, Siberia, Canada, Alaska, China and Bangladesh. An estimated 200 million hectares of Gleysols are found in the tropics, mainly in the Amazon region, equatorial Africa and the coastal swamps of Southeast Asia.

A stagnohumic gley soil in a forest plantation in mid-Wales, U.K. The organic-rich topsoil is over a grey and orange mottled subsoil developed in glacial till ("boulder clay").

Gley soils are grouped under Gleysols in the FAO World Reference Base for Soil Resources. They exhibit a greenish-blue-grey soil color due to anoxic wetland conditions. On exposure, as the iron in the soil oxidizes colors are transformed to a mottled pattern of reddish, yellow or orange patches. During soil formation (*gleying*), the oxygen supply in the soil profile is restricted due to soil moisture at saturation. Anaerobic micro-organisms support cellular respiration by using alternatives to free oxygen as electron acceptors to support cellular respiration. Where Anaerobic organisms reduce ferric oxide to ferrous oxide, the reduced mineral compounds produce the gley soil typical color. Green rust, a layered double hydroxide (LDH) of Fe(II) and Fe(III) can be found as the mineral fougerite in gley soils.

Gley soils may be sticky and hard to work, especially where the gleying is caused by surface water, held up on a slowly permeable layer. However, some ground-water gley soils have permeable lower horizons, including some sands, for example in hollows within sand dune systems, known as slacks, and in some alluvial situations. Groundwater gley soils develop where drainage is poor because the water table (phreatic surface) is high, whilst Surface-water gleying occurs when precipitation inputs at the surface do not drain freely through the ground. A reducing environment exists in the saturated layers, which become mottled greyish-blue or brown because of the content of ferrous iron and organic matter. The presence of reddish or orange mottles indicates localised re-oxidation of ferrous salts in the soil matrix, and is often associated with root channels, animal burrows or cracking of the soil material during dry spells.

Gypsisols

Gypsisols in the FAO World Reference Base for Soil Resources are soils with substantial secondary accumulation of gypsum ($CaSO_4.2H_2O$). They are found in the driest parts of the arid climate zone.In the USDA soil taxonomy they are classified as Gypsids (USDA Soil Taxonomy), in the Russian soil classification they are called *Desert soils* (USSR).

Calcic Gypsisol found in the US.

Gypsisols are developed in mostly unconsolidated alluvial, colluvial and aeolian deposits of base-rich weathering material. They are found on level to hilly land in arid regions. The natural vegetation is sparse and dominated by xerophyticshrubs and trees and/or ephemeral grasses

These soils have ABC profiles. Accumulation of calcium sulphate, with or without carbonates, is concentrated in and below the B-horizon.

Deep Gypsisols located close to water resources can be planted to a wide range of crops. Yields are severely depressed where a petrogypsic horizon occurs at shallow depth. Nutrient imbalance, stoniness, and uneven subsidence of the land surface upon dissolution of gypsum in percolating (irrigation) water are further limitations. Irrigation canals must be lined to prevent the canal walls from caving in. Large areas of Gypsisols are in use for low volume grazing. Gypsisols are exclusive to arid regions; their worldwide extent is probably of the order of 100 million hectares. Major occurrences are in and around Mesopotamia in desert areas in the Middle East and adjacent central Asian republics, in the Libyan and Namib deserts, in southeast and central Australia and in the southwestern United States.

REFERENCES

1. Jackson, J. A. (1997). Glossary of Geology (4. ed.). Alexandria, Virginia: American Geological Institute. p 604. ISBN 0-922152-34-9
2. H. H. Janzen; et al. (2011). "Global Prospects Rooted in Soil Science". Soil Science Society of America Journal. 75 (1): 1. doi:10.2136/sssaj2009.0216.
3. Pielke, Roger (December 12, 2005). "Is Soil an Important Component of the Climate System?". The Climate Science Weblog. Archived from the original on 8 September 2006. Retrieved 19 April 2012.
4. "Vernacular Systems". Archived from the original on 6 March 2007. Retrieved 19 April 2012.
5. Krasilnikov, N.A. (1958) Soil Microorganisms and Higher Plants
6. Wikisource:The New Student's Reference Work/4-0310
7. Buol, S. W.; Hole, F. D. & McCracken, R. J. (1973). Soil Genesis and Classification (First ed.). Ames, IA: Iowa State University Press. ISBN 978-0-8138-1460-5..
8. C. C. Nikiforoff (1959). "Reappraisal of the soil: Pedogenesis consists of transactions in matter and energy between the soil and its surroundings". Science. 129 (3343): 186–196. Bibcode:1959Sci...129..186N. doi:10.1126/science.129.3343.186. PMID 17808687.

CHAPTER 7

Ecological Studies of Soil

Chapter Contents

The gut contents of isopods consisted mainly of fragmentary tissues of needles and subsidiarity of fungal hyphae. The proportion of fungi in the gut varied with the season of the year, the age of isopods and their habitats in relation to fungal density.

The more the decay of needles proceeded, the more the needles were preferred and consumed, *i.e.* brown needles least, dark brown needles intermediate, and blackish brown and yellowish decayed needles most. Irrespective of considerable differences in food consumption of these needles, assimilation was the same and its efficiency decreased with the progress of decay.

As far as the needles in the incipient stages of decay are concerned, the structural toughness of needles is considered to be the prominent factor in determining the feeding value of the needles. Of extractive treatments on natural needles, removals of ether- and water-soluble fractions in all kinds of needles reduced drastically both consumption and assimilation.

SOIL ECOLOGY

Soil ecology is the study of the interactions among soil organisms, and between biotic and abiotic aspects of the soil environment. It is particularly concerned with the cycling of nutrients, formation and stabilization of the pore structure, the spread and vitality of pathogens, and the biodiversity of this rich biological community. Soil is made up of a multitude of physical, chemical, and biological entities, with many interactions occurring among them. Soil is a variable mixture of broken and weathered minerals and decaying organic matter. Together with the proper amounts of air and water, it supplies, in part, sustenance for plants as well as mechanical support.

The diversity and abundance of soil life exceeds that of any other ecosystem. Plant establishment, competitiveness, and growth is governed largely by the ecology below-ground, so understanding this system is an essential component of plant sciences and terrestrial ecology.

Features of the Ecosystem

- Moisture is a major limiting factor on land. Terrestrial organisms are constantly confronted with the problem of dehydration. Transpiration or evaporation of water from plant surfaces is an energy dissipating process unique to the terrestrial environment.
- Temperature variations and extremes are more pronounced in the air than in the water medium.
- On the other hand, the rapid circulation of air throughout the globe results in a ready mixing and remarkably constant content of oxygen and carbon dioxide.
- Although soil offers solid support, air does not. Storing skeletons have been evolved in both land plants and animals and also special means of locomotion have been evolved in the latter.
- Land, unlike the ocean, is not continuous; there are important geographical barriers to free movement.
- The nature of the substrate, although important in water is especially vital in terrestrial environment. Soil, not air, is the source of highly variable nutrients; it is a highly developed ecological subsystem.

Soil Food Web

An incredible diversity of organisms make up the soil food web. They range in size from the tiniest one-celled bacteria, algae, fungi, and protozoa, to the more complex nematodes and micro-arthropods, to the visible earthworms, insects, small vertebrates, and plants. As these organisms eat, grow, and move through the soil, they make it possible to have clean water, clean air, healthy plants, and moderated water flow.

There are many ways that the soil food web is an integral part of landscape processes. Soil organisms decompose organic compounds, including manure, plant residues, and pesticides, preventing them from entering water and becoming pollutants. They sequester nitrogen and other nutrients that might otherwise enter groundwater, and they fix nitrogen from the atmosphere, making it available to plants. Many organisms enhance soil aggregation and porosity, thus increasing infiltration and reducing surface runoff. Soil organisms prey on crop pests and are food for above-ground animals.

Research

Research interests span many aspects of soil ecology and microbiology, Fundamentally, researchers are interested in understanding the interplay among microorganisms, fauna, and plants, the biogeochemical processes they carry out, and the physical environment in which their activities take place, and applying this knowledge to address environmental problems.

Example research projects are to examine the biogeochemistry and microbial ecology of septic drain field soils used to treat domestic wastewater, the role of anecic earthworms in controlling the movement of water and nitrogen cycle in agricultural soils, and the assessment of soil quality in turf production.

Of particular interest as of 2006 is to understand the roles and functions of arbuscular mycorrhizal fungi in natural ecosystems. The effect of anthropic soil conditions on arbuscular mycorrhizal fungi, and the production of glomalin by arbuscular mycorrhizal fungi are both of particular interest due to their roles in sequestering atmospheric carbon dioxide.

EDAPHOLOGY

Edaphology is one of two main divisions of soil science, the other being pedology. Edaphology is concerned with the influence of soils on living things, particularly plants. The term is also applied to the study of how soil influences humankind's use of land for plant growth as well as man's overall use of the land. General subfields within edaphology are agricultural soil science (known by the term agrology in some regions) and environmental soil science. (Pedology deals with pedogenesis, soil morphology, and soil classification.)

In Russia, edaphology is considered equivalent to pedology, but is recognized to have an applied sense consistent with agrophysics and agrochemistry outside Russia.

Xenophon (431–355 BC), and Cato (234–149 BC), were early edaphologists. Xenophon noted the beneficial effect of turning a cover crop into the earth. Cato wrote De Agri Cultura (On Farming) which recommended tillage, crop rotation and the use of legumes in the rotation to build soil nitrogen. He also devised the first soil capability classification for specific crops.

Jan Baptist van Helmont (1577–1644) performed a famous experiment, growing a willow tree in a pot of soil and supplying only rainwater for five years. The weight gained by the tree was greater than the weight loss of the soil. He concluded that the willow was made of water. Although only partly correct, his experiment reignited interest in edaphology.

Areas of Study

Agricultural Soil Science

Agricultural soil science is the application of soil chemistry, physics, and biology dealing with the production of crops. In terms of soil chemistry, it places particular emphasis on plant nutrients of importance to farming and horticulture, especially with regard to soil fertility and fertilizer components.

Physical edaphology is strongly associated with crop irrigation and drainage.

Soil husbandry is a strong tradition within agricultural soil science. Beyond preventing soil erosion and degradation in cropland, soil husbandry seeks to sustain the agricultural soil resource though the use of soil conditioners and cover crops.

Environmental Soil Science

Environmental soil science studies our interaction with the pedosphere on beyond crop production. Fundamental and applied aspects of the field address vadose zone functions, septic drain field site assessment and function, land treatment of wastewater, stormwater, erosion control, soil contamination with metals and pesticides, remediation of contaminated soils, restoration of wetlands, soil degradation, and environmental nutrient management. It also studies soil in the context of land use planning, global warming, and acid rain.

AGROECOLOGY

Agroecology is the study of ecological processes applied to agricultural production systems. The prefix *agro-* refers to *agriculture*. Bringing ecological principles to bear in agroecosystems can suggest novel management approaches that would not otherwise be considered. The term is often used imprecisely and may refer to "a science, a movement, [or] a practice". Agroecologists study a variety of agroecosystems, and the field of agroecology is not associated with any one particular method of farming, whether it be organic, integrated, or conventional; intensive or extensive. Although it has much more common thinking and principles with some of the before mentioned farming systems.

A community-supported agriculture share of crops.

Ecological Strategy

Agroecologists do not unanimously oppose technology or inputs in agriculture but instead assess how, when, and if technology can be used in conjunction with natural, social and human assets. Agroecology proposes a context- or site-specific manner of studying agroecosystems, and as such, it recognizes that there is no universal formula or recipe for the success and maximum well-being of an agroecosystem. Thus, agroecology is not defined by certain management practices, such as the use of natural enemies in place of insecticides, or polyculture in place of monoculture.

Instead, agroecologists may study questions related to the four system properties of agroecosystems: productivity, stability, sustainability and equitability. As opposed to disciplines that are concerned with only one or some of these properties, agroecologists see all four properties as interconnected and integral to the success of an agroecosystem. Recognizing that these properties are found on varying spatial scales, agroecologists do not limit themselves to the study of agroecosystems at any one scale: gene-organism-population-community-ecosystem-landscape-biome, field-farm-community-region-state-country-continent-global.

Agroecologists study these four properties through an interdisciplinary lens, using natural sciences to understand elements of agroecosystems such as soil properties and plant-insect interactions, as well as using social sciences to understand the effects of farming practices on rural communities, economic constraints to developing new production methods, or cultural factors determining farming practices.

Approaches

Agroecologists do not always agree about what agroecology is or should be in the long-term. Different definitions of the term agroecology can be distinguished largely by the specificity with which one defines the term "ecology", as well as the term's potential political connotations. Definitions of agroecology, therefore, may be first grouped according to the specific contexts within which they situate agriculture. Agroecology is defined by the OECD as "the study of the relation of agricultural crops and environment." This definition refers to the "-ecology" part of "agroecology" narrowly as the natural environment. Following this definition, an agroecologist would study agriculture's various relationships with soil health, water quality, air quality, meso- and micro-fauna, surrounding flora, environmental toxins, and other environmental contexts.

A more common definition of the word can be taken from Dalgaard et al., who refer to agroecology as the study of the interactions between plants, animals, humans and the environment within agricultural systems. Consequently, agroecology is inherently multidisciplinary, including factors from agronomy, ecology, sociology, economics and related disciplines. In this case, the "-ecology" portion of "agroecology is defined broadly to include social, cultural, and economic contexts as well. Francis et al. also expand the definition in the same way, but put more emphasis on the notion of food systems.

Agroecology is also defined differently according to geographic location. In the global south, the term often carries overtly political connotations. Such political definitions of the term usually ascribe to it the goals of social and economic justice; special attention, in this case, is often paid to the traditional farming knowledge of indigenous populations. North American and European uses of the term sometimes avoid the inclusion of such overtly political goals. In these cases, agroecology is seen more strictly as a scientific discipline with less specific social goals.

Agro-Population Ecology

This approach is derived from the science of ecology primarily based on population ecology, which over the past three decades has been displacing the ecosystems biology of Odum. Buttel explains the main difference between the two categories, saying that "the application of population ecology to agroecology involves the primacy not only of analyzing agroecosystems from the perspective of the population dynamics of their constituent species, and their relationships to climate and biogeochemistry, but also there is a major emphasis placed on the role of genetics."

Inclusive Agroecology

Rather than viewing agroecology as a subset of agriculture, Wojtkowski takes a more encompassing perspective. In this, natural ecology and agroecology are the major headings under ecology. Natural ecology is the study of organisms as they interact with and within natural environments. Correspondingly, agroecology is the basis for the land-use sciences. Here humans are the primary governing force for organisms within planned and managed, mostly terrestrial, environments.

As key headings, natural ecology and agroecology provide the theoretical base for their respective sciences. These theoretical bases overlap but differ in a major way. Economics has no role in the functioning of natural ecosystems whereas economics sets direction and purpose in agroecology.

Under agroecology are the three land-use sciences, agriculture, forestry, and agroforestry. Although these use their plant components in different ways, they share the same theoretical core.

Beyond this, the land-use sciences further subdivide. The subheadings include agronomy, organic farming, traditional agriculture, permaculture, and silviculture. Within this system of subdivisions, agroecology is philosophically neutral. The importance lies in providing a theoretical base hitherto lacking in the land-use sciences. This allows progress in biocomplex agroecosystems including the multi-species plantations of forestry and agroforestry.

APPLICATIONS

To arrive at a point of view about a particular way of farming, an agroecologist would first seek to understand the contexts in which the farm(s) is(are) involved. Each farm may be inserted in a unique combination of factors or contexts. Each farmer may have their own premises about the meanings of an agricultural endeavor, and these meanings might be different from those of agroecologists. Generally, farmers seek a configuration that is viable in multiple contexts, such as family, financial, technical, political, logistical, market, environmental, spiritual. Agroecologists want to understand the behavior of those who seek livelihoods from plant and animal increase, acknowledging the organization and planning that is required to run a farm.

Views on Organic and Non-Organic Milk Production

Because organic agriculture proclaims to sustain the health of soils, ecosystems, and people, it has much in common with Agroecology; this does not mean that Agroecology is synonymous with organic agriculture, nor that Agroecology views organic farming as the ‹right› way of farming. Also, it is important to point out that there are large differences in organic standards among countries and certifying agencies.

Three of the main areas that agroecologists would look at in farms, would be: the environmental impacts, animal welfare issues, and the social aspects.

Environmental impacts caused by organic and non-organic milk production can vary significantly. For both cases, there are positive and negative environmental consequences.

Compared to conventional milk production, organic milk production tends to have lower eutrophication potential per ton of milk or per hectare of farmland, because it potentially reduces leaching of nitrates (NO_3^-) and phosphates (PO_4^-) due to lower fertilizer application rates. Because organic milk production reduces pesticides utilization, it increases land use per ton of milk due to decreased crop yields per hectare. Mainly due to the lower level of concentrates given to cows in organic herds, organic dairy farms generally produce less milk per cow than conventional dairy farms. Because of the increased use of roughage and the, on-average, lower milk production level per cow, some research has connected organic milk production with increases in the emission of methane.

Animal welfare issues vary among dairy farms and are not necessarily related to the way of producing milk (organically or conventionally).

A key component of animal welfare is freedom to perform their innate (natural) behavior, and this is stated in one of the basic principles of organic agriculture. Also, there are other aspects of animal welfare to be considered – such as freedom from hunger, thirst, discomfort, injury, fear, distress, disease and pain. Because organic standards require loose housing systems, adequate bedding, restrictions on the area of slatted floors, a minimum forage proportion in the ruminant diets, and tend to limit stocking densities both on pasture and in housing for dairy cows, they potentially promote good foot and hoof health. Some studies show lower incidence of placenta retention, milk fever, abomasums displacement and other diseases in organic than in conventional dairy herds. However, the level of infections by parasites in organically managed herds is generally higher than in conventional herds.

Social aspects of dairy enterprises include life quality of farmers, of farm labor, of rural and urban communities, and also includes public health.

Both organic and non-organic farms can have good and bad implications for the life quality of all the different people involved in that food chain. Issues like labor conditions, labor hours and labor rights, for instance, do not depend on the organic/non-organic characteristic of the farm; they can be more related to the socio-economical and cultural situations in which the farm is inserted, instead.

As for the public health or food safety concern, organic foods are intended to be healthy, free of contaminations and free from agents that could cause human diseases. Organic milk is meant to have no chemical residues to consumers, and the restrictions on the use of antibiotics and chemicals in organic food production has the purpose to accomplish this goal. Although dairy cows in both organic and conventional farming practices can be exposed to pathogens, it has been shown that, because antibiotics are not permitted as a preventative measure in organic practices, there are far fewer antibiotic resistant pathogens on organic farms. This dramatically increases the efficacy of antibiotics when/if they are necessary.

In an organic dairy farm, an agroecologist could evaluate the following:

- Can the farm minimize environmental impacts and increase its level of sustainability, for instance by efficiently increasing the productivity of the animals to minimize waste of feed and of land use?
- Are there ways to improve the health status of the herd (in the case of organics, by using biological controls, for instance)?
- Does this way of farming sustain good quality of life for the farmers, their families, rural labor and communities involved?

Views on no-till Farming

No-tillage is one of the components of conservation agriculture practices and is considered more environmental friendly than complete tillage. There is a general consensus that no-till can increase soils capacity of acting as a carbon sink, especially when combined with cover crops.

No-till can contribute to higher soil organic matter and organic carbon content in soils, though reports of no-effects of no-tillage in organic matter and organic carbon soil contents also exist, depending on environmental and crop conditions. In addition, no-till can indirectly reduce CO_2 emissions by decreasing the use of fossil fuels.

Most crops can benefit from the practice of no-till, but not all crops are suitable for complete no-till agriculture. Crops that do not perform well when competing with other plants that grow in untilled soil in their early stages can be best grown by using other conservation tillage practices, like a combination of strip-till with no-till areas. Also, crops which harvestable portion grows underground can have better results with strip-tillage, mainly in soils which are hard for plant roots to penetrate into deeper layers to access water and nutrients.

The benefits provided by no-tillage to predators may lead to larger predator populations, which is a good way to control pests (biological control), but also can facilitate predation of the crop itself. In corn crops, for instance, predation by caterpillars can be higher in no-till than in conventional tillage fields.

In places with rigorous winter, untilled soil can take longer to warm and dry in spring, which may delay planting to less ideal dates. Another factor to be considered is that organic residue from the prior year›s crops lying on the surface of untilled fields can provide a favorable environment to pathogens, helping to increase the risk of transmitting diseases to the future crop. And because no-till farming provides good environment for pathogens, insects and weeds, it can lead farmers to a more intensive use of chemicals for pest control. Other disadvantages of no-till include underground rot, low soil temperatures and high moisture.

Based on the balance of these factors, and because each farm has different problems, agroecologists will not atest that only no-till or complete tillage is the right way of farming. Yet, these are not the only possible choices regarding soil preparation, since there are intermediate practices such as strip-till, mulch-till and ridge-till, all of them – just as no-till – categorized as conservation tillage. Agroecologists, then, will evaluate the need of different practices for the contexts in which each farm is inserted.

REFERENCES

1. IPCC Special Report on Land Use, Land-Use Change And Forestry, 2.2.1.1 Land Use
2. FAO Land and Water Division retrieved 14 September 2010
3. JAPA 25:3
4. Village of Euclid, Ohio v. Ambler Realty Co.
5. Nolon, John R., Local Land Use Control in New York: An Aging Citadel Under Siege (July/Aug. 1992). New York State Bar Journal, p. 38, July–August 1992.
6. Zoning
7. Ameztegui, Aitor; Coll, Lluis; Brotons, Lluis; Ninot, JM (2016). "Land-use legacies rather than climate change are driving the recent upward shift of the mountain tree line in the Pyrenees." (PDF). Global Ecology and Biogeography. doi:10.1111/geb.12407.
8. Robert Hart (1996). Forest Gardening. p. 124. Forest gardening, in the sense of finding uses for and attempting to control the growth of wild plants, is undoubtedly the oldest form of land use in the world.
9. [dead link]Intergovernmental Panel on Climate Change
10. UN Land Degradation and Land Use/Cover Data Sources ret. 26 June 2007
11. UN Report on Climate Change retrieved 25 June 2007 from Web archive
12. Jaeker WG, Plantinga AJ (2007). How have Land-use regulations Affected Property Values in Oregon? OSU Extension.
13. Johnson, David E. (2008). "4. Marketing Studies and Market Considerations". Fundamentals of Land Development. Hoboken, NJ: John Wiley & Sons. ISBN 9780471778936.
14. Lynch, Kevin (1960). The Image of the City. Cambridge MA: MIT Press. OL 5795447M.
15. Top soil is removed from the new road bed (Video). Dayton, TN: Bryan College. 2009-09-03.

16. Urban Soil Compaction (PDF) (Report). Soil Quality – Urban Technical Note No. 2. Auburn, AL: U.S. Natural Resources Conservation Service, Soil Quality Instsitute. March 2000.
17. Wilson, Sarah Jane (2014-08-14). "Not All Forests Are Created Equal: Reforesting the Tropics for People, Biodiversity, and Carbon". The Equation (Blog). Cambridge, MA: Union of Concerned Scientists.
18. Boucher, Doug (June 2011). The Root of the Problem: What's Driving tropical Deforestation today? (PDF) (Report). Tropical Forest and Climate Initiative. Cambridge, MA: Union of Concerned Scientists. p. 85.
19. "Monocultures". Barcelona, Spain: Carbon Trade Watch. Retrieved 2016-08-28.
20. Brockerhoff, E.G.; Jactel, H.; Parrotta, J.; Quine, C.P.; Sayer, J. (2008-04-09). "Plantation forests and biodiversity: oxymoron or opportunity?". Biodiversity and Conservation. Springer. 17 (5): 925–951. doi:10.1007/s10531-008-9380-x.
21. R.J. Oosterbaan, International Institute for Land Reclamation and Improvement, Wageningen, The Netherlands. "Improvement of waterlogged and saline soils." Free downloads of software and articles on land drainage.
22. Kone, D. Linda (2006). Land Development (10th ed.). Washington, D.C.: National Association of Home Builders. ISBN 9780867186093.
23. Dewberry & Davis (2008). Land Development Handbook (3rd ed.). New York: McGraw-Hill Professional. ISBN 9780071640930.

CHAPTER 8

Soil Biology

Chapter Contents

While you work long hours to produce tons of hay or corn from each acre of land, tons of bacteria, fungi, insects, and other organisms are working underground, making farming possible. They decompose organic matter and transform nutrients into forms your crops can use. They help build good soil tilth, enhance crop growth, and control pests.

INTRODUCTION

Soil biology is the study of microbial and faunal activity and ecology in soil. Soil life, soil biota, soil fauna, or edaphon is a collective term that encompasses all the organisms that spend a significant portion of their life cycle within a soil profile, or at the soil-litter interface. These organisms include earthworms, nematodes, protozoa, fungi, bacteria and different arthropods. Soil biology plays a vital role in determining many soil characteristics. The decomposition of organic matter by soil organisms has an immense influence on soil fertility, plant growth, soil structure, and carbon storage. As a relatively new science, much remains unknown about soil biology and their effects on soil ecosystems.

The soil is home to a large proportion of the world's biodiversity. The links between soil organisms and soil functions are observed to be incredibly complex. The interconnectedness and complexity of this soil 'food web' means any appraisal of soil function must necessarily take into account interactions with the living communities that exist within the soil. We know that soil organisms break down organic matter, making nutrients available for uptake by plants and other organisms. The nutrients stored in the bodies of soil organisms prevent nutrient loss by leaching. Microbial exudates act to maintain soil structure, and earthworms are important in bioturbation. However, we find that we don't understand critical aspects about how these populations function and interact. The discovery of glomalin in 1995 indicates that we lack the knowledge to correctly answer some of the most basic questions about the biogeochemical cycle in soils. We have much work ahead to gain a better understanding of how soil biological components affect us and the biosphere

In balanced soil, plants grow in an active and steady environment. The mineral content of the soil and its heartiful structure are important for their well-being, but it is the life in the earth that powers its cycles and provides its fertility. Without the activities of soil organisms, organic materials would accumulate and litter the soil surface, and there would be no food for plants. The soil biota includes:

- Megafauna: size range - 20 mm upward, e.g. moles, rabbits, and rodents.
- macrofauna: size range - 2 to 20 mm, e.g. woodlice, earthworms, beetles, centipedes, slugs, snails, ants, and harvestmen.
- Mesofauna: size range - 100 micrometres to 2 mm, e.g. tardigrades, mites and springtails.
- Microfauna and Microflora: size range - 1 to 100 micrometres, e.g. yeasts, bacteria (commonly actinobacteria), fungi, protozoa, roundworms, and rotifers.

Of these, bacteria and fungi play key roles in maintaining a healthy soil. They act as decomposers that break down organic materials to produce detritus and other breakdown products. Soil detritivores, like earthworms, ingest detritus and decompose it. Saprotrophs, well represented by fungi and bacteria, extract soluble nutrients from delitro. The ants (macrofaunas) help by breaking down in the same way but they also provide the motion part as they move in their armies. Also the rodents, wood-eaters help the soil to be more absorbent.

SCOPE

Soil biology involves work in the following areas:

- Modelling of biological processes and population dynamics
- Soil biology, physics and chemistry: occurrence of physicochemical parameters and surface properties on biological processes and population behavior
- Population biology and molecular ecology: methodological development and contribution to study microbial and faunal populations; diversity and population dynamics; genetic transfers, influence of environmental factors
- Community ecology and functioning processes: interactions between organisms and mineral or organic compounds; involvement of such interactions in soil pathogenicity; transformation of mineral and organic compounds, cycling of elements; soil structuration

Complementary disciplinary approaches are necessarily utilized which involve molecular biology, genetics, ecophysiology, biogeography, ecology, soil processes, organic matter, nutrient dynamics and landscape ecology.

Bacteria

Bacteria are single-cell organisms and the most numerous denizens of agriculture, with populations ranging from 100 million to 3 billion in a gram. They are capable of very rapid reproduction by binary fission (dividing into two) in favourable conditions. One bacterium is capable of producing 16 million more in just 24 hours. Most soil bacteria live close to plant roots and are often referred to as rhizobacteria. Bacteria live in soil water, including the film of moisture surrounding soil particles, and some are able to swim by means of flagella. The majority of the beneficial soil-dwelling bacteria need oxygen (and are thus termed aerobic bacteria), whilst those that do not require air are referred to as anaerobic, and tend to cause putrefaction of dead organic matter. Aerobic bacteria are most active in a soil that is moist (but not saturated, as this will deprive aerobic bacteria of the air that they require), and neutral soil pH, and where there is plenty of food (carbohydrates and micronutrients from organic matter) available. Hostile conditions will not completely kill bacteria; rather, the bacteria will stop growing and get into a dormant stage, and those individuals with pro-adaptive mutations may compete better in the new conditions. Some gram-positive bacteria produce spores in order to wait for more favourable circumstances, and gram-negative bacteria get into a "nonculturable" stage. Bacteria are colonized by persistent viral agents (bacteriophages) that determine gene word order in bacterial host.

From the organic gardener's point of view, the important roles that bacteria play are:

Nitrification

Nitrification is a vital part of the nitrogen cycle, wherein certain bacteria (which manufacture their own carbohydrate supply without using the process of photosynthesis) are able to transform nitrogen in the form of ammonium, which is produced by the decomposition of proteins, into nitrates, which are available to growing plants, and once again converted to proteins.

Nitrogen Fixation

In another part of the cycle, the process of nitrogen fixation constantly puts additional nitrogen into biological circulation. This is carried out by free-living nitrogen-fixing bacteria in the soil or water such as *Azotobacter*, or by those that live in close symbiosis with leguminous plants, such as rhizobia. These bacteria form colonies in nodules they create on the roots of peas, beans, and related species. These are able to convert nitrogen from the atmosphere into nitrogen-containing organic substances.

Denitrification

While nitrogen fixation converts nitrogen from the atmosphere into organic compounds, a series of processes called denitrification returns an approximately equal amount of nitrogen to the atmosphere. Denitrifying bacteria tend to be anaerobes, or facultatively anaerobes (can alter between the oxygen dependent and oxygen independent types of metabolisms), including *Achromobacter* and *Pseudomonas*. The purification process caused by oxygen-free conditions converts nitrates and nitrites in soil into nitrogen gas or into gaseous compounds such as nitrous oxide or nitric oxide. In excess, denitrification can lead to overall losses of available soil nitrogen and subsequent loss of soil fertility. However, fixed nitrogen may circulate many times between organisms and the soil before denitrification returns it to the atmosphere. The diagram above illustrates the nitrogen cycle.

Actinobacteria

Actinobacteria are critical in the decomposition of organic matter and in humus formation, and their presence is responsible for the sweet "earthy" aroma associated with a good healthy soil. They require plenty of air and a pH between 6.0 and 7.5, but are more tolerant of dry conditions than most other bacteria and fungi.

Fungi

A gram of garden soil can contain around one million fungi, such as yeasts and moulds. Fungi have no chlorophyll, and are not able to photosynthesise. They cannot use atmospheric carbon dioxide as a source of carbon, therefore they are chemo-heterotrophic, meaning that, like animals, they require a chemical source of energy rather than being able to use light as an energy source, as well as organic substrates to get carbon for growth and development.

Many fungi are parasitic, often causing disease to their living host plant, although some have beneficial relationships with living plants, as illustrated below. In terms of soil and humus creation, the most important fungi tend to be saprotrophic; that is, they live on dead or decaying organic matter, thus breaking it down and converting it to forms that are available to the higher plants. A succession of fungi species will colonise the dead matter, beginning with those that use sugars and starches, which are succeeded by those that are able to break down cellulose and lignins.

Fungi spread underground by sending long thin threads known as mycelium throughout the soil; these threads can be observed throughout many soils and compost heaps. From the mycelia the fungi is able to throw up its fruiting bodies, the visible part above the soil (e.g., mushrooms, toadstools, and puffballs), which may contain millions of spores. When the fruiting body bursts, these spores are dispersed through the air to settle in fresh environments, and are able to lie dormant for up to years until the right conditions for their activation arise or the right food is made available.

Mycorrhizae

Those fungi that are able to live symbiotically with living plants, creating a relationship that is beneficial to both, are known as Mycorrhizae (from *myco* meaning fungal and *rhiza* meaning root). Plant root hairs are invaded by the mycelia of the mycorrhiza, which lives partly in the soil and partly in the root, and may either cover the length of the root hair as a sheath or be concentrated around its tip. The mycorrhiza obtains the carbohydrates that it requires from the root, in return providing the plant with nutrients including nitrogen and moisture. Later the plant roots will also absorb the mycelium into its own tissues. Beneficial mycorrhizal associations are to be found in many of our edible and flowering crops. Shewell Cooper suggests that these include at least 80% of the *brassica* and *solanum* families (including tomatoes and potatoes), as well as the majority of tree species, especially in forest and woodlands. Here the mycorrhizae create a fine underground mesh that extends greatly beyond the limits of the tree's roots, greatly increasing their feeding range and actually causing neighbouring trees to become physically interconnected. The benefits of mycorrhizal relations to their plant partners are not limited to nutrients, but can be essential for plant reproduction: In situations where little light is able to reach the forest floor, such as the North American pine forests, a young seedling cannot obtain sufficient light to photosynthesise for itself and will not grow properly in a sterile soil. But, if the ground is underlain by a mycorrhizal mat, then the developing seedling will throw down roots that can link with the fungal threads and through them obtain the nutrients it needs, often indirectly obtained from its parents or neighbouring trees.

David Attenborough points out the plant, fungi, animal relationship that creates a Three way harmonious trio to be found in forest ecosystems, wherein the plant/fungi symbiosis is enhanced by animals such as the wild boar, deer, mice, or flying squirrel, which feed upon the fungi›s fruiting bodies, including truffles, and cause their further spread (*Private Life Of Plants*, 1995). A greater understanding of the complex relationships that pervade natural systems is one of the major justifications of the organic gardener, in refraining from the use of artificial chemicals and the damage these might cause.

Recent research has shown that arbuscular mycorrhizal fungi produce glomalin, a protein that binds soil particles and stores both carbon and nitrogen. These glomalin-related soil proteins are an important part of soil organic matter.

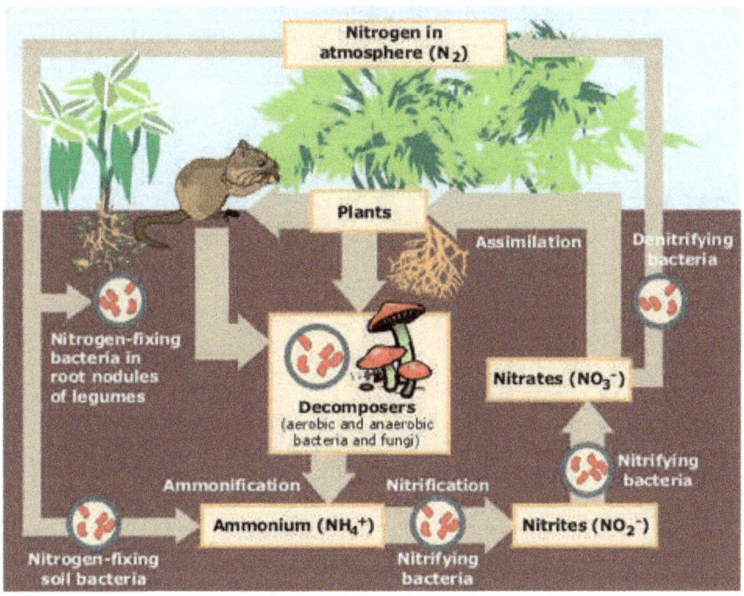

The nitrogen cycle (Turkish)

Earthworms, Ants, and Termites

Earthworms, ants and termites mix the soil as they burrow, significantly affecting soil formation. Earthworms ingest soil particles and organic residues, enhancing the availability of plant nutrients in the material that passes through and out their bodies. By aerating and stirring the soil, and by increasing the stability of soil aggregates, these organisms help assure the ready infiltration of water.

REFERENCES

1. Jackson, J. A. (1997). Glossary of Geology (4. ed.). Alexandria, Virginia: American Geological Institute. p 604. ISBN 0-922152-34-9
2. H. H. Janzen; et al. (2011). "Global Prospects Rooted in Soil Science". Soil Science Society of America Journal. 75 (1): 1. doi:10.2136/sssaj2009.0216.
3. Pielke, Roger (December 12, 2005). "Is Soil an Important Component of the Climate System?". The Climate Science Weblog. Archived from the original on 8 September 2006. Retrieved 19 April 2012.
4. "Vernacular Systems". Archived from the original on 6 March 2007. Retrieved 19 April 2012.
5. Krasilnikov, N.A. (1958) Soil Microorganisms and Higher Plants
6. Wikisource:The New Student's Reference Work/4-0310
7. Buol, S. W.; Hole, F. D. & McCracken, R. J. (1973). Soil Genesis and Classification (First ed.). Ames, IA: Iowa State University Press. ISBN 978-0-8138-1460-5..
8. C. C. Nikiforoff (1959). "Reappraisal of the soil: Pedogenesis consists of transactions in matter and energy between the soil and its surroundings". Science. 129 (3343): 186–196. Bibcode:1959Sci...129..186N. doi:10.1126/science.129.3343.186. PMID 17808687.

CHAPTER 9

Soil Conservation and Management

Chapter Contents

Soils are essential for the continuity of life on earth since many ecosystems depend on them for survival. Still, soils persist to face serious threats in so many ways. Soil erosion, use of chemical pesticides and herbicides, unsustainable agricultural practices, excessive farming, water pollution and land pollution are some of the aspects upsetting the natural functions of the soil. This is where soil conservation comes in.

Soil conservation is the practice of protecting the soil against erosion or deterioration. It involves the activities that can be undertaken to ensure our soils are at their optimum quality and health.

INTRODUCTION

Soil conservation is the prevention of soil loss from erosion or reduced fertility caused by over usage, acidification, salinization or other chemical soil contamination.

Slash-and-burn and other unsustainable methods of subsistence farming are practiced in some lesser developed areas. A sequel to the deforestation is typically large scale erosion, loss of soil nutrients and sometimes total desertification.

Erosion barriers on disturbed slope, Marin County, California

Techniques for improved soil conservation include crop rotation, cover crops, conservation tillage and planted windbreaks and affect both erosion and fertility. When plants, especially trees, die, they decay and become part of the soil. Code 330 defines standard methods recommended by the US Natural Resources Conservation Service.

Contour plowing, Pennsylvania 1938. The rows formed slow water run-off during rainstorms to prevent soil erosion and allows the water time to infiltrate into the soil.

Farmers have practiced soil conservation for millennia. Conservation practices fall in multiple categories:

CONTOUR PLOWING

Contour plowing orients crop furrows following the contour lines of the farmed area. Furrows move left and right to maintain a constant altitude, which reduces runoff. Contour plowing was practiced by the ancient Phoenicians, and is effective for slopes between two and ten percent. Contour plowing can increase crop yields from 10 to 50 percent, partially as a result from greater soil retention.

Terracing or terrace farming

Terracing is the practice of creating nearly level areas in a hillside area. The terraces form a series of steps, each at a higher level than the previous. Terraces are protected from erosion by other soil barriers.

Terraced farming is more common on small farms and in underdeveloped countries, since mechanized equipment is difficult to deploy in this setting. It protects the soil from its erosion. It is one of the way by which soil erosion can be stopped.

KEYLINE DESIGN

Keyline design is an enhancement of contour farming, where the total watershed properties are taken into account in forming the contour lines.

PERIMETER RUNOFF CONTROL

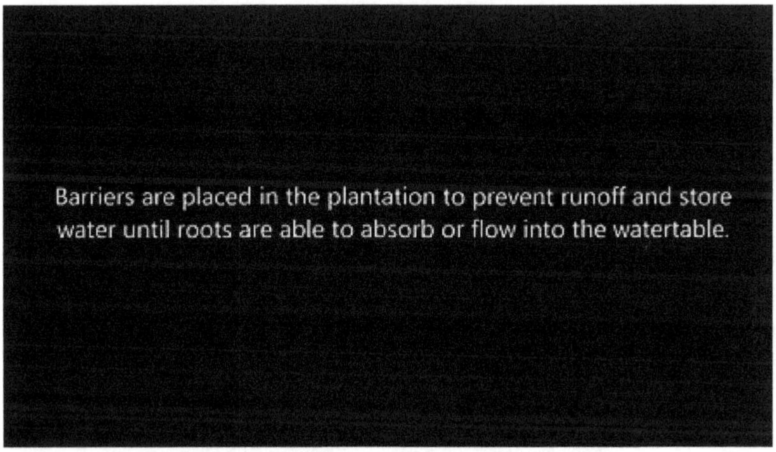

runoff and filter soxx

Tree, shrubs and ground-cover are effective perimeter treatment for soil erosion prevention, by impeding surface flows. A special form of this perimeter or inter-row treatment is the use of a “grass way” that both channels and dissipates runoff through surface friction, impeding surface runoff and encouraging infiltration of the slowed surface water.

WINDBREAKS

Windbreaks are sufficiently dense rows of trees at the windward exposure of an agricultural field subject to wind erosion. Evergreen species provide year-round protection; however, as long as foliage is present in the seasons of bare soil surfaces, the effect of deciduous trees may be adequate.

COVER CROPS/CROP ROTATION

Cover crops such as legumes, white turnip, radishes and other species are rotated with cash crops to blanket the soil year-round and act as green manure that replenishes nitrogen and other critical nutrients. Cover crops also help suppress weeds.

SOIL-CONSERVATION FARMING

Soil-conservation farming involves no-till farming, "green manures" and other soil-enhancing practices. Such farming methods attempt to mimic the biology of virgin land. They can revive damaged soil, minimize erosion, encourage plant growth, eliminate the use of nitrogen fertilizer or fungicide, produce above-average yields and protect crops during droughts or flooding. The result is less labor and lower costs that increase farmers' profits. No-till farming and cover crops act as sinks for nitrogen and other nutrients. This increases the amount of soil organic matter.

Repeated plowing/tilling degrades soil, killing its beneficial fungi and earthworms. Once damaged, soil may take multiple seasons to fully recover, even in optimal circumstances.

Critics argue that no-till and related methods are impractical and too expensive for many growers, partly because it requires new equipment. They cite advantages for conventional tilling depending on the geography, crops and soil conditions. Some farmers claimed that no-till complicates weed control, delays planting and that post-harvest residues, especially for corn, are hard to manage.

Salinity Management

Salt deposits on the former bed of the Aral Sea

Salinity in soil is caused by irrigating with salty water. Water then evaporates from the soil leaving the salt behind. Salt breaks down the soil structure, causing infertility and reduced growth.

The ions responsible for salination are: sodium (Na^{+}), potassium (K^{+}), calcium (Ca^{2+}), magnesium (Mg^{2+}) and chlorine (Cl^{-}). Salinity is estimated to affect about one third of the earth's arable land. Soil salinity adversely affects crop metabolism and erosion usually follows. Salinity occurs on drylands from overirrigation and in areas with shallow saline water tables. Over-irrigation deposits salts in upper soil layers as a byproduct of soil infiltration; irrigation merely increases the rate of salt deposition. The best-known case of shallow saline water table capillary action occurred in Egypt after the 1970 construction of the Aswan Dam. The change in the groundwater level led to high salt concentrations in the water table. The continuous high level of the water table led to soil salination.

Use of humic acids may prevent excess salination, especially given excessive irrigation. Humic acids can fix both anions and cations and eliminate them from root zones. Planting species that can tolerate saline conditions to produce surface cover can preserve soil salinity can be reduced. Salt-tolerant plants include saltbush, a plant found in much of North America and in the Mediterranean regions of Europe.

SOIL ORGANISMS

Yellow fungus, a mushroom that assists in organic decay.

When worms excrete egesta in the form of casts, a balanced selection of minerals and plant nutrients is made into a form accessible for root uptake. Earthworm casts are five times richer in available nitrogen, seven times richer in available phosphates and eleven times richer in available potash than the surrounding upper 150 millimetres (5.9 in) of soil. The weight of casts produced may be greater than 4.5 kg per worm per year. By burrowing, the earthworm improves soil porosity, creating channels that enhance the processes of aeration and drainage.

Other important soil organisms include nematodes, mycorrizha and bacteria.

Degraded soil requires synthetic fertilizer to produce high yields. Lacking structure increases erosion and carries nitrogen and other pollutants into rivers and streams. Each one percent increase in soil organic matter helps soil hold **20,000** gallons more water per acre.

Mineralization

To allow plants full realization of their phytonutrient potential, active mineralization of the soil is sometimes undertaken. This can involve adding crushed rock or chemical soil supplements. In either case the purpose is to combat mineral depletion. A broad range of minerals can be used, including common substances such as phosphorus and more exotic substances such as zinc and selenium. Extensive research examines the phase transitions of minerals in soil with aqueous contact.

Flooding can bring significant sediment to an alluvial plain. While this effect may not be desirable if floods endanger life or if the sediment originates from productive land, this process of addition to a floodplain is a natural process that can rejuvenate soil chemistry through mineralization.

SOIL MANAGEMENT

Soil management concerns all operations, practices, and treatments used to protect soil and enhance its performance (such as soil fertility or soil mechanics). It includes soil conservation, soil amendment, and optimal soil health. In agriculture, some amount of soil management is needed both in nonorganic and organic types to prevent agricultural land from becoming poorly productive over decades. Organic farming in particular stresses optimal soil management, because it uses soil health as the exclusive or nearly exclusive source of its fertilization and pest control.

Environment

According to the EPA, agricultural soil management practices can lead to production and emission of nitrous oxide (N_2O), a major greenhouse gas and air pollutant. Activities that can contribute to N2O emissions include fertilizer usage, irrigation and tillage. The management of soils accounts for over half of the emissions from the Agriculture sector. Cattle livestocks account for one third of emissions, through methane emissions. Manure management and rice cultivation also emit emissions.

Methods that significantly enhance carbon sequestration in soil include no-till farming, residue mulching, cover cropping, and crop rotation, all of which are more widely used in organic farming than in conventional farming. Because only 5% of US farmland currently uses no-till and residue mulching, there is a large potential for carbon sequestration.

Practices

Soil management practices that affect soil quality:

- Controlling traffic on the soil surface helps to reduce soil compaction, which can reduce aeration and water infiltration.
- Cover crops keep the soil anchored and covered in off-seasons so that the soil is not eroded by wind and rain.
- Crop rotations for row crops alternate high-residue crops with lower-residue

crops to increase the amount of plant material left on the surface of the soil during the year to protect the soil from erosion.

- Nutrient management can help to improve the fertility of the soil and the amount of organic matter content, which improves soil structure and function.
- Tillage, especially reduced-tillage or no-till operations limit the amount of soil disturbance while cultivating a new crop, and help to maintain plant residues on the surface of the soil for erosion protection and water retention.

ADVANTAGES OF SOIL MANAGEMENT

- Maintain soil fertility
- Restore soil fertilitys
- Make the agricultural process an economic one
- Help increase yield

REFERENCES

1. Predicting soil erosion by water, a guide to conservation planning in the Revised Universal Soil Loss Equation, United States Department of Agriculture, Agricultural Research Service, Agricultural handbook no. 703 (1997)
2. Contour farming boosts yields : a farmer's guide in laying out key contour lines and establishing grassed waterways. United States. 1943. p. 1. Retrieved 1 August 2015. |first1= missing |last1= in Authors list (help)
3. Perimeter landscaping of Carneros Business Park, Lumina Technologies, Santa Rosa, Ca., prepared for Sonoma County, Ca. (2002)
4. Wolfgang Summer, Modelling Soil Erosion, Sediment Transport and Closely Related Hydrological Processes entry by Mingyuan Du, Peiming Du, Taichi Maki and Shigeto Kawashima, "Numerical modeling of air flow over complex terrain concerning wind erosion", International Association of Hydrological Sciences publication no. 249 (1998) ISBN 1-901502-50-3
5. Goode, Erica (March 10, 2015). Farmers Put Down the Plow for More Productive Soil. The New York Times (New York ed.). The New York Times Company. p. D1. ISSN 0362-4331. OCLC 1645522. Retrieved April 5, 2015.
6. Dan Yaron, Salinity in Irrigation and Water Resources, Marcel Dekker, New York (1981) ISBN 0-8247-6741-1
7. Bill Mollison, Permaculture: A Designer›s Manual, Tagari Press, (December 1, 1988), 576 pages, ISBN 0908228015. Increases in porosity enhance infiltration and thus reduce adverse effects of surface runoff.
8. Arthur T. Hubbard, Encyclopedia of Surface and Colloid Science Vol 3, Santa Barbara, California Science Project, Marcel Dekker, New York (2004) ISBN 0-8247-0759-1
9. Agriculture: Sources of Greenhouse Gas Emissions. EPA. 2015.

10. Susan S. Lang (13 July 2005). Organic farming produces same corn and soybean yields as conventional farms, but consumes less energy and no pesticides, study finds. Retrieved 8 July 2008.
11. Pimentel, David; Hepperly, Paul; Hanson, James; Douds, David; Seidel, Rita (2005). Environmental, Energetic, and Economic Comparisons of Organic and Conventional Farming Systems. BioScience. 55 (7): 573–82. doi:10.1641/0006-3568(2005)055[0573:EEAECO]2.0.CO;2.
12. Lal, Rattan; Griffin, Michael; Apt, Jay; Lave, Lester; Morgan, M. Granger (2004). Ecology: Managing Soil Carbon. Science. 304 (5669): 393. doi:10.1126/science.1093079. PMID 15087532.
13. Soilquality.org - Soil Management Practices
14. Soil Management - Penn State

CHAPTER 10

Land Management

Chapter Contents

Land management practices describe the way that land is managed - the means by which a land use outcome is achieved.

Land management practices have become an increasing focus of ACLUMP because the drivers for better land use outcomes (economic, social and/or environmental) inevitably come from how land is managed. An understanding of the links between land use, land management practice and resource condition supports on-ground decisions to change land management practice, and to support sustainable land use policies and programs.

INTRODUCTION

Land use involves the management and modification of natural environment or wilderness into built environment such as settlements and semi-natural habitats such as arable fields, pastures, and managed woods. It also has been defined as «the total of arrangements, activities, and inputs that people undertake in a certain land cover type.»

Habitat fragmented by numerous roads near the Indiana Dunes National Lakeshore

REGULATION

Land use practices vary considerably across the world. The United Nations' Food and Agriculture Organization Water Development Division explains that "Land use concerns the products and/or benefits obtained from use of the land as well as the land management actions (activities) carried out by humans to produce those products and benefits." As of the early 1990s, about 13% of the Earth was considered arable land, with 26% in pasture, 32% forests and woodland, and 1.5% urban areas.

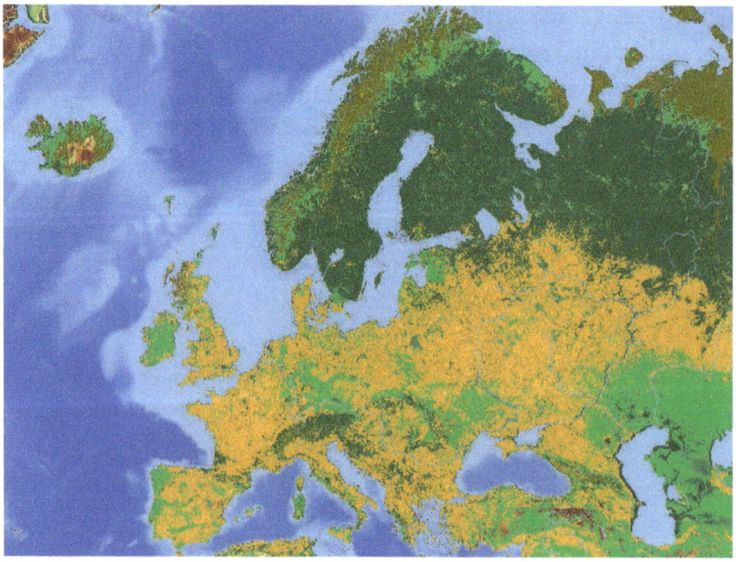

A land use map of Europe—major non-natural land uses include arable farmland (yellow) and pasture (light green).

As Albert Guttenberg (1959) wrote many years ago, "'Land use' is a key term in the language of city planning." Commonly, political jurisdictions will undertake land-use planning and regulate the use of land in an attempt to avoid land-use conflicts. Land use plans are implemented through land division and use ordinances and regulations, such as zoning regulations. Management consulting firms and non-governmental organizations will frequently seek to influence these regulations before they are codified.

United States

In colonial America, few regulations existed to control the use of land, due to the seemingly endless amounts of it. As society shifted from rural to urban, public land regulation became important, especially to city governments trying to control industry, commerce, and housing within their boundaries. The first zoning ordinance was passed in New York City in 1916, and, by the 1930s, most states had adopted zoning laws. In the 1970s, concerns about the environment and historic preservation led to further regulation.

Today, federal, state, and local governments regulate growth and development through statutory law. The majority of controls on land, however, stem from the actions of private developers and individuals. Three typical situations bringing such private entities into the court system are: suits brought by one neighbor against another; suits brought by a public official against a neighboring landowner on behalf of the public; and suits involving individuals who share ownership of a particular parcel of land. In these situations, judicial decisions and enforcement of private land-use arrangements can reinforce public regulation, and achieve forms and levels of control that regulatory zoning cannot.

Two major federal laws have been passed in the last half century that limit the use of land significantly. These are the National Historic Preservation Act of 1966 (today embodied in 16 U.S.C. 461 et seq.) and the National Environmental Policy Act of 1969 (42 U.S.C. 4321 et seq.).

ENVIRONMENT

Land use and land management practices have a major impact on natural resources including water, soil, nutrients, plants and animals. Land use information can be used to develop solutions for natural resource management issues such as salinity and water quality. For instance, water bodies in a region that has been deforested or having erosion will have different water quality than those in areas that are forested. Forest gardening, a plant-based food production system, is believed to be the oldest form of land use in the world.

The major effect of land use on land cover since 1750 has been deforestation of temperate regions. More recent significant effects of land use include urban sprawl, soil erosion, soil degradation, salinization, and desertification. Land-use change, together with use of fossil fuels, are the major anthropogenic sources of carbon dioxide, a dominant greenhouse gas.

According to a report by the United Nations' Food and Agriculture Organisation, land degradation has been exacerbated where there has been an absence of any land use planning, or of its orderly execution, or the existence of financial or legal incentives that have led to the wrong land use decisions, or one-sided central planning leading to over-utilization of the land resources - for instance for immediate production at all costs. As a consequence the result has often been misery for large segments of the local population and destruction of valuable ecosystems. Such narrow approaches should be replaced by a technique for the planning and management of land resources that is integrated and holistic and where land users are central. This will ensure the long-term quality of the land for human use, the prevention or resolution of social conflicts related to land use, and the conservation of ecosystems of high biodiversity value.

The citadel of Kastellet, Copenhagen that has been converted into a park, showing multiple examples of suburban land use

URBAN GROWTH BOUNDARIES

The urban growth boundary is one form of land-use regulation. For example, Portland, Oregon is required to have an urban growth boundary which contains at least 20,000 acres (81 km^2) of vacant land. Additionally, Oregon restricts the development of farmland. The regulations are controversial, but an economic analysis concluded that farmland appreciated similarly to the other land.

LAND DEVELOPMENT

Land development refers to altering the landscape in any number of ways such as:

- Changing landforms from a natural or semi-natural state for a purpose such as agriculture or housing
- Subdividing real estate into lots, typically for the purpose of building homes
- Real estate development or changing its purpose, for example by converting an unused factory complex into condominia.

Since their invention, heavy equipment such as bulldozers have been useful for earthmoving in land development.

ECONOMIC ASPECTS

In an economics context, land development is also sometimes advertised as land improvement or land amelioration. It refers to investments making land more usable by humans. For accounting purposes it refers to any variety of projects that increase the value of the property. Most are depreciable, but some land improvements are not able to be depreciated because a useful life cannot be determined. Home building and containment are two of the most common and the oldest types of development.

In an urban context, land development furthermore includes:

- Road construction
  - Access roads, walkways and parking lots
  - Bridging
- Landscaping
  - Clearing, terracing or land levelling
- Setup of fences and, to a lesser degree, hedges
- Service connections to municipal services and public utilities
- Drainage, canals

- External lighting (street lamps, etc.)

Landowner or developers on any size of project will often want to maximise profits, minimise risk and control cash flow. This "profit enhancement" means identifying and developing the best scheme for the local marketplace, whilst satisfying the local planning process.

Development Analysis puts development prospects and the development process itself under the microscope, identifying where enhancements and improvements can be introduced. These improvements aim to align with best design practice, political sensitivities, and the inevitable social requirements of a project, with the overarching objective of increasing land values and profit margins on behalf of the landowner or developer.

Stonefields Suburb Development

a

b

c

Development analysis can add significantly to the value of land and development, and as such is a crucial tool for landowners and developers. It is an essential step in Kevin A. Lynch's 1960 book *The Image of the City*, and is considered to be essential to realizing the value potential of land. The landowner can share in additional planning gain (significant value uplift) via an awareness of the land's development potential. This is done via a residual development appraisal or residual valuation. The residual appraisal calculates the sale value of the end product (the Gross Development Value, or GDV), and hypothetically deducts costs, including planning and construction costs, finance costs and developer's profit. The "residue", or leftover proportion, represents the land value. Therefore, in maximising the GDV (that which one could build on the land), land value is concurrently enhanced.

Land value is highly sensitive to supply and demand (for the end product), build costs, planning and affordable housing contributions, and so on. Understanding the intricacies of the development system and the effect of "value drivers" can result in massive differences in the landowner's sale value.

Conversion of Landforms

Land development puts more emphasis on the expected economic development as a result of the process; "land conversion" tries to focus on the general physical and biological aspects of the land use change. "Land improvement" in the economic sense can often lead to land degradation from the ecological perspective. Land development and the change in land value does not usually take into account changes in the ecology of the developed area. While conversion of (rural) land with a vegetation carpet to building land may result in a rise in economic growth and rising land prices, the irreversibility of lost flora and fauna because of habitat destruction, the loss of ecosystem services and resulting decline in environmental value is only considered *a priori* in environmental full-cost accounting.

Conversion to Building Land

Motorway construction in Ireland

Conversion to building land is as a rule associated with road building, which in itself already brings topsoil abrasion, soil compaction and modification of the soil's chemical composition through soil stabilization, creation of impervious surfaces and, subsequently, (polluted) surface runoff water.

Building Construction

Construction activity often effectively seals off a larger part of the soil from rainfall and the nutrient cycle, so that the soil below buildings and roads is effectively "consumed" and made infertile.

With the notable exception of attempts at rooftop gardening and hanging gardens in green buildings (possibly as constituents of green urbanism), vegetative cover of higher plants is lost to concrete and asphalt surfaces, complementary interspersed garden and park areas notwithstanding.

Conversion to Farmland

In Argentina and Bolivia, the Chacothorn forest (A) is being felled at a rate considered among the highest in the world (B), to give way to soybean cultivation (C)

New creation of farmland will rely on the conversion and development of previous forests, savannas or grassland. Recreation of farmland from wasteland, deserts or previous impervious surfaces is considerably less frequent because of the degraded or missing fertile soil in the latter. Starting from forests, land is made arable by assarting or slash-and-burn. Agricultural development furthermore includes:

- Hydrological measures (land levelling, drainage, irrigation, sometimes landslide and flood control)
- Soil improvement (fertilization, establishment of a productive chemical balance).
- Road construction

Oil palm plantation and rainforest fragment on Borneo

Because the newly created farmland is more prone to erosion than soil stabilized by tree roots, such a conversion may mean irreversible crossing of an ecological threshold.

Horton overland flow

The resulting deforestation is also not easily compensated for by reforestation or afforestation. This is because plantations of other trees as a means for water conservation and protection against wind erosion (shelterbelts), as a rule, lack the biodiversity of the lost forest, especially when realized as monocultures. Also, soil stabilization and erosion control measures may not be as effective in preserving topsoil as the previous intact vegetation.

Restoration

Massive land conversion without proper consideration of ecological and geological consequences may lead to disastrous results, such as:

- General soil degradation
  - Catastrophic soil salination and solonchak formation, e.g., in Central Asia, as a consequence of irrigation by saline groundwater
  - Desertification, soil erosion and ecological shifts due to drainage
  - Leaching of saline soils
- Habitat loss for the wildlife.

While deleterious effects can be particularly visible when land is developed for industrial or mining usage, agro-industrial and settlement use can also have a massive and sometimes irreversible impact on the affected ecosystem.

Examples of land restoration/land rehabilitation counted as land development in the strict sense are still rare. However, renaturation, reforestation, stream restoration may all contribute to a healthier environment and quality of life, especially in densely populated regions. The same is true for planned vegetation like parks and gardens, but restoration plays a particular role, because it reverses previous conversions to built and agricultural areas.

LAND MANAGEMENT

Land management is the process of managing the use and development (in both urban and rural settings) of land resources. Land resources are used for a variety of purposes which may include organic agriculture, reforestation, water resource management and eco-tourism projects.

RFERENCES

1. IPCC Special Report on Land Use, Land-Use Change And Forestry, 2.2.1.1 Land Use
2. FAO Land and Water Division retrieved 14 September 2010
3. JAPA 25:3
4. Village of Euclid, Ohio v. Ambler Realty Co.
5. Nolon, John R., Local Land Use Control in New York: An Aging Citadel Under Siege (July/Aug. 1992). New York State Bar Journal, p. 38, July–August 1992.
6. Zoning
7. Ameztegui, Aitor; Coll, Lluis; Brotons, Lluis; Ninot, JM (2016). "Land-use legacies rather than climate change are driving the recent upward shift of the mountain tree line in the Pyrenees." (PDF). Global Ecology and Biogeography. doi:10.1111/geb.12407.

8. Robert Hart (1996). Forest Gardening. p. 124. Forest gardening, in the sense of finding uses for and attempting to control the growth of wild plants, is undoubtedly the oldest form of land use in the world.
9. [dead link]Intergovernmental Panel on Climate Change
10. UN Land Degradation and Land Use/Cover Data Sources ret. 26 June 2007
11. UN Report on Climate Change retrieved 25 June 2007 from Web archive
12. Jaeker WG, Plantinga AJ (2007). How have Land-use regulations Affected Property Values in Oregon? OSU Extension.
13. Johnson, David E. (2008). "4. Marketing Studies and Market Considerations". Fundamentals of Land Development. Hoboken, NJ: John Wiley & Sons. ISBN 9780471778936.
14. Lynch, Kevin (1960). The Image of the City. Cambridge MA: MIT Press. OL 5795447M.
15. Top soil is removed from the new road bed (Video). Dayton, TN: Bryan College. 2009-09-03.
16. Urban Soil Compaction (PDF) (Report). Soil Quality – Urban Technical Note No. 2. Auburn, AL: U.S. Natural Resources Conservation Service, Soil Quality Instsitute. March 2000.
17. Wilson, Sarah Jane (2014-08-14). "Not All Forests Are Created Equal: Reforesting the Tropics for People, Biodiversity, and Carbon". The Equation (Blog). Cambridge, MA: Union of Concerned Scientists.
18. Boucher, Doug (June 2011). The Root of the Problem: What's Driving tropical Deforestation today? (PDF) (Report). Tropical Forest and Climate Initiative. Cambridge, MA: Union of Concerned Scientists. p. 85.
19. "Monocultures". Barcelona, Spain: Carbon Trade Watch. Retrieved 2016-08-28.
20. Brockerhoff, E.G.; Jactel, H.; Parrotta, J.; Quine, C.P.; Sayer, J. (2008-04-09). "Plantation forests and biodiversity: oxymoron or opportunity?". Biodiversity and Conservation. Springer. 17 (5): 925–951. doi:10.1007/s10531-008-9380-x.
21. R.J. Oosterbaan, International Institute for Land Reclamation and Improvement, Wageningen, The Netherlands. "Improvement of waterlogged and saline soils." Free downloads of software and articles on land drainage.
22. Kone, D. Linda (2006). Land Development (10th ed.). Washington, D.C.: National Association of Home Builders. ISBN 9780867186093.
23. Dewberry & Davis (2008). Land Development Handbook (3rd ed.). New York: McGraw-Hill Professional. ISBN 9780071640930.

CHAPTER 11

Soil Governance

Chapter Contents

The implementation of the SDGs depends on natural resources, including land and soil. The sustainable use of these resources has to be integrated for success. Current governance arrangements and the degree to which they are implemented are, however, far from being appropriate to respond to the land-related and soil-related challenges to sustainable development. Among other things, governance approaches need to better address power imbalances. Appropriate accountability mechanisms within the post-2015 Development Agenda would be one way of addressing this challenge.

INTRODUCTION

Soil governance refers to the policies, strategies, and the processes of decision-making employed by nation states and local governments regarding the use of soil. Globally, governance of the soil has been limited to an agricultural perspective due to increased food insecurity from the most populated regions on earth. The Global Soil Partnership, GSP, was initiated by the Food and Agriculture Organization (FAO) and its members with the hope to improve governance of the limited soil resources of the planet in order to guarantee healthy and productive soils for a food-secure world, as well as support other essential ecosystem services. Governing the soil requires international and national collaboration between governments, local authorities, industries and citizens to ensure implementation of coherent policies that encourage practices and methodologies that regulate usage of the resource to avoid conflict between users to promote sustainable land management. In the European Union›s environmental policies, soil is recognized as a non-renewable resource, but its governance is maintained at a national level, unlike other non-renewable and climate sensitive resources. In the developing world, soil governance is biased towards promoting sustainable agriculture and ensuring food security.

Governance of the soil differs from soil management. Soil management involves practices and techniques used to increase and maintain soil fertility, structure, and carbon sequestration, etc. Soil management techniques are heavily utilized in agriculture, because of the need to regulate the various practices, such as tillage techniques, fertilizer application and crop rotation (among others) by the various stakeholders involved. The need to monitor and avoid the negative effects of agricultural land use such as soil erosion has formed the basis of the discourse and awareness on soil governance, and has also seen the emergence of science and technology as the link between soil management and governance. Soil governance mechanisms are usually encapsulated within the context of land governance, with little focus on urban and industrial soil governance especially in developing countries that have rapid urbanization rates; thus, soil governance is highly interlinked with other atmospheric and anthropogenic processes which may contribute to the difficulty in distinguishing it as an entity.

With an aim to make soil data available to all, the Food and Agriculture Organization and UNESCO created a global soil map in 1981 as the main information on the distribution of soil resources. Currently, under the GSP framework, a new global soil information system will be developed.

In 2002, the International Union of Soil Sciences proposed December 5 to be «World Soil Day» to celebrate the importance of soil in our lives. Under the framework of the GSP, the sixty-eighth session of the United Nations General Assembly in December 2013 designated December 5 as the World Soil Day and declared 2015 as the International Year of Soils with the aim to raise awareness on the importance of soils for ecosystem functions and food security .

GLOBAL SOIL PARTNERSHIP

Changes in land use, population growth, and the impacts of climate change have led to a gradual process of soil degradation. Soil degradation is a gradual process involving the natural and anthropogenic processes that result in the physical loss (erosion) and reduction in soil quality. The recognition of anthropogenic effects on soil degradation has influenced discourse of urban soil management, and formulation of policies by regional organizations. However, soil remains as the primary medium for food production, thus global soil governance is directed towards the impacts of soil degradation on food production and conflicts that arise between the need for human settlements and space available for food production. The impacts of climate change also contribute to the conflict because carbon dioxide emissions have progressively led to higher average global temperatures, which has led to an increase in soil degradation through erosion, increased salinity, and a reduction in the flora and fauna that contribute to its quality.

The Global Soil Partnership (GSP) was launched in 2011. It is an initiative of the FAO and the European Commission that aims to "Support and facilitate joint efforts towards sustainable management of soil resources for food security and climate change adaptation and mitigation" and to update the 1982 World Soil Charter. The World Conference on Agrarian Reform and Rural Development (Rome, 1979), and the Plan of Action of the UN Conference on Desertification (Nairobi, 1977) to combat land degradation and desertification initiated international discourse on effective soil management and governance, however the initiatives were limited to the soil and its relation to agricultural productivity.

The GSP›s ultimate goal is to achieve food security and restoration of ecosystem services through conserving, enhancing, and restoring soil resources through productive and sustainable use. In addition to being a global partnership, the GSP aims to create Regional Soil Partnerships (RSPs) to provide guidance on goals and priorities within specific regions and develop relevant activities within each region.

The GSP meets in Plenary Assembly form to meet the different demands from each region. The First Plenary Assembly, held in June 2013 at the FAO›s headquarters adopted the Rules of Procedure, nominated and established an Intergovernmental Technical Panel of Soils (ITPS), started thinking about the 5 pillars of action, supported the implementation of Regional Soil Partnerships and developed a GSP roadmap. The Second Plenary Assembly will be held in July 2014.

Vision and Mission

The vision of the GSP is to improve governance of soil resources to guarantee healthy and productive soils for a food-secure world and support other essential ecosystem services.

The mission of the GSP is capacity building, facilitate and contribute to soil science and technologies for sustainable management of soil resources at all levels.

Intergovernmental Panel of Soils (ITPS)

The Intergovernmental Panel of Soils represents all regions of the world and is composed of 27 soil expert representatives. The ITPS is here to advise the GSP on scientific and technical knowledge, advocate for the inclusion of sustainable soil management in different agendas, review the GSP's Plan of Action, follow up on the Plan of Action and request the formation of committees for exceptional cases.

Five Pillars of Action

To achieve its vision and mandate, the GSP operates under five pillars of action:

1. Promote sustainable management of soil resources for soil protection, conservation and sustainable productivity.
2. Encourage investment, technical cooperation, policy, education awareness and extension in soil
3. Promote targeted soil research and development focusing on identified gaps and priorities and synergies with related productive, environmental and social development actions

4. Enhance the quantity and quality of soil data and information: data collection (generation), analysis, validation, reporting, monitoring and integration with other disciplines
5. Harmonization of methods, measurements and indicators for the sustainable management and protection of soil resources

INTERNATIONAL YEAR OF SOILS

Logo of International Year of Soils 2015

The International Year of Soils, 2015 (IYS 2015) was declared by the Sixty-eighth session of the United Nations General Assembly on December 20, 2013 after recognizing December 5 as World Soil Day.

Food and Agriculture Organization of the United Nations has been nominated to implement the IYS 2015, within the framework of the Global Soil Partnership and in collaboration with Governments and the secretariat of the United Nations Convention to Combat Desertification.

The aim of the IYS is to increase awareness and understanding of the importance of soil for food security and essential ecosystem functions.

IYS Objectives

The International Year of Soils aims to:

- Raise full awareness among civil society and decision makers about the profound importance of soil for human life;
- Educate the public about the crucial role soil plays in food security, climate change adaptation and mitigation, essential ecosystem services, poverty alleviation and sustainable development;
- Support effective policies and actions for the sustainable management and protection of soil resources;
- Promote investment in sustainable soil management activities to develop and maintain healthy soils for different land users and population groups;
- Strengthen initiatives in connection with the SDG process (Sustainable Development Goals) and Post-2015 agenda;

Advocate for rapid capacity enhancement for soil information collection and monitoring at all levels (global, regional and national).

BY REGION

In the European Union

The United Nations Earth Summit Conference on Environment and Development catalysed the formation of EU environmental policy into a policy that was centred on the environmental consequences of integration. The conference organized by the United Nations saw the acceptance of various documents and charters governing the natural environment and sustainable development. In its inception phase, EU environmental policy was a reaction to normalise competition with the markets. Having a common policy would ensure that member states were bound to directives that would regulate their production methods thus affect production output and competitive advantage. Focus was directed on air pollution from industries, and other forms of tangible, measurable, and traceable pollution that could be isolated to an event or a process, such as acidification Swedish lakes in the 1970s and 1980s caused by high sulphur dioxide emissions from power plants. Such forms of pollution were often governed with bans, quotas and economic instruments such as taxes and fines. With improvements in technology, access and delivery of information and changes in global conceptualisation of the environment, EU environmental policy has evolved to become more responsive and bespoke, and has also increased its scope to various sources and sinks of pollution.

In 2006, the European Commission tabled a proposal to the European Parliament to establish a framework for soil protection within the union. Soil is recognised as a non-renewable resource because of its slow formation process. However—unlike other non-renewable resources, such as coal, that have explicit policies governing extraction methods, trading, and consumption in the EU—soil governance is contained within the contexts of environmental policies and regulation on various entities of the biosphere. The draft policy recognised that soil governance had been "scattered" in EU legislation, and lacked a cohesive isolated framework, therefore governance and management of the same resource was open to interpretation depending on the main resource and industry policy in question. The policy sought to unify the "scattered" regulations because they lacked the mandate to "identify and cover all soil threats". This view was supported by extensive consultation between stakeholders and the European commission that started in February 2003, and saw member states convey their support for a framework based on regional action in 2004. The framework was developed as a directive to member states; this form of legislation allows interpretation by stakeholders at national and local levels, and between networks thus complying with the subsidiarity principle. The principle provides that EU political decisions must be implemented at "the lowest possible administrative and political level, and as close to the citizens as possible", unless in areas where action by individual countries is insufficient. It is under this principle that member states rejected the proposal to establish a framework for soil protection as the proposal argued that member states are unable to effectively monitor and manage their soils. Inconsistencies in national soil governance strategies and, classification and treatment of contaminants would disable the objectives of the proposal because of the complexities of trans-border soil pollution and management. Furthermore, soil degradation and mismanagement affects other environmental areas, and industries governed through EU legislation such as water, biodiversity, and food production, thus it was deemed appropriate to have uniform legislation across all entities. Member states argued that soil management should not be negotiated at the European Regional level as they already had strong domestic policies regulating soil usage and management, therefore focus should be directed at strengthening local policies and regulatory institution. Consequently, the EU does not have a cohesive soil governance policy and relies on environmental policies, and non-renewable resource policies and legislations of member states to guide utilisation, management and regulate pollutants of the soil.

In India

In terms of employment, the agricultural sector is one of the main sectors in the Indian economy. In 2010, the sector employed 58.2% of the nations workforce, and contributed 15.7% to the nations GDP. Cognisant of agriculture's role in the economy, the 11th five-year economic plan that runs from 2007-2012 recognises the importance of proper soil management in agriculture. Soil degradation through excessive and miscalculated fertiliser use because of emphasis on increased output has led to nearly two-thirds of India's farmlands to be classified as either degraded or sick. In attempts to increase knowledge on soils and soil management, the government of Gujarat initiated the Soil Health Cards Programme in 2006 that was "expected to bridge the gap between Scientists, agricultural extension workers, farmers and input-output dealers". The programme relies on technology to disseminate accountable and uncomplicated scientific information that is based on the farmers needs. Farmers take samples of their soil for analysis in a state run laboratory. Based on the sample, farmers get information on the soil mineral and water content, fertiliser application methods, and advice on what crops to grow.

In the pilot scheme, collected data was inputted into a web-based information system that included the Internet, intranet and GSWAN (Gujarat State Wide Area Network) to build up the state, and national database on soil health. Increasing knowledge on soil management, increased output, and reduced costs for farmers and contributed to Gujarat's agriculture growth rate that was three times the national growth rate in 2009. The success of the scheme has facilitated its implementation at a national level under the Ministry of Agriculture. Each state and union territory is responsible for the set-up and management of the soil testing facilities and maintaining the state soil database so that it is uniform and standardised. The testing, advisory and issuance processes involved are at multi-governance levels involving stakeholders from the private and public sectors. Government approved NGOs, community associations' farmers, the state administration and national administration are all involved in the scheme at different levels. The process begins with the farmers assisted by various NGOs and community groups and involves interactions with more NGOs and state officers at higher levels, as they are responsible for soil sample testing. The impacts of increased global temperatures have had negative effects towards effective soil management techniques in the developing world. Changes in precipitation patterns and an increase in extreme events such as floods and drought have exacerbated issues such as desertification and soil erosion. The effects of such events are further aggravated by resource-deficient farmers and government officials who lack skills to prepare and manage their soils for disasters, and end up relying on relief aid for sustenance. Addressing the Impacts of desertification is a complex multi-level process because of the social, environmental and economic factors.

The Republic of India ratified the United Nations Convention to Combat Desertification in 1984, and has since instituted the National Programme to Combat Desertification that utilises an integrated and holistic approach at governing the soil. Through various ministries the national programme aspires to implement the UNCCD by increasing capacity in soil and soil water management, improving access to micro-credit with a focus on women and marginalised groups, promoting alternative energy supplies to reduce reliance on wood increasing soil monitoring and strengthening legislation regarding land management in industrial and mining activities.

IMPACTS OF INDUSTRIALIZATION AND URBANIZATION

In the developing world, industrialization and urbanization are emerging as significant contributors to land and soil degradation. Lack of sufficient knowledge in soil management and disregard for the environment has been identified as key reasons affecting urban soil degradation. Industrialization alters the chemical aspects of the soil through pollution of heavy metals and effluents. Construction and landfills in urban areas affect soil through compaction and excavation, which affect natural processes such as water purification and storage. In the developing and developing world governance of soils in urban areas requires bespoke policies because of the nature of urban and industrial developments in the cities. In Central Europe, governance of urban soils is facilitated by the urban soil management strategy that aims to design applicable soil management strategies in select European cities. Through networks created with universities and municipal authorities the project aims to research into and develop an interdisciplinary approach to manage urban soils,. In the developing world, urban and industrial soil governance is linked to sustainable development of cities to address urban poverty and responsible land use through effective waste management. Often, developing countries lack resources to implement policies governing settlements and industries, thus the soil and water are often heavily polluted. Urban soil management uses an interdisciplinary approach to protect biodiversity reliant upon the soil and land, reduce pollution from industrial effluents, and increase resilience of the soil to stresses such as compaction from construction.

ROLE OF SCIENCE AND TECHNOLOGY

The global soil map is a global consortium between academic, regional and national scientific institutions coordinated by stakeholders according to the respective regions. Using Geographic Information Systems (GIS), Remote sensing and emerging technologies a global soil map will be created to represent different soil types. The consortium is led by the ISRIC-World Soil Information, whose mandate to increase knowledge on soils through data collection and dissemination and development of technologies and methodologies for digital soil mapping. GIS is used to display, analyse and collate soil data and processes, and also identify different types of soils through mapping and web-based software. Soil Science is used in tandem with GIS to identify individual soil properties applicable to agricultural and urban soil management. The Soil Health Card Programme in India utilises soil science to advise farmers on fertiliser usage and crop rotations and records the data on a national network which can be used to map different soil types across the country.

CRITICISMS

Exponential population growth has led to discourses on maintaining food security through increased agricultural output. Often emphasis is limited to increasing intensity of output through more intensive use of fertilisers, machinery, pesticides and other inputs. Because soil is the main medium in agriculture, soil governance has maintained a strong agricultural perspective. Urban soil governance is crucial in developing countries because of rapid urbanization and changing land use in rural areas. Industrial activities such as mining dramatically alter the landscape and soil through generation of overburden and excavation; however governance remains biased towards agriculture and food production. The effects of climate change vis-à-vis desertification and increased soil erosion are not yet addressed in cohesive global policies, yet the scope and nature of the problem is global. The establishment of the GSP in 2011 has recognized the need for an interdisciplinary approach to soil governance underpinning the diverse ecosystem services and functions including the production of healthy food.

SOIL TEST

Soil test may refer to one or more of a wide variety of soil analyses conducted for one of several possible reasons. Possibly the most widely conducted soil tests are those done to estimate the plant-available concentrations of plant nutrients, in order to determine fertilizer recommendations in agriculture. Other soil tests may be done for engineering (geotechnical), geochemical or ecological investigations.

Plant Nutrition

In agriculture, a **soil test** commonly refers to the analysis of a soil sample to determine nutrient content, composition, and other characteristics such as the acidity or pH level. A soil test can determine fertility, or the expected growth potential of the soil which indicates nutrient deficiencies, potential toxicities from excessive fertility and inhibitions from the presence of non-essential trace minerals. The test is used to mimic the function of roots to assimilate minerals. The expected rate of growth is modeled by the Law of the Maximum.

Labs, such as those at Iowa State and Colorado State University, recommend that a soil test contains 10-20 sample points for every 40 acres (160,000 m^2) of field. Tap water or chemicals can change the composition of the soil, and may need to be tested separately. As soil nutrients vary with depth and soil components change with time, the depth and timing of a sample may also affect results.

Composite sampling can be performed by combining soil from several locations prior to analysis. This is a common procedure, but should be used judiciously to avoid skewing results. This procedure must be done so that government sampling requirements are met. A reference map should be created to record the location and quantity of field samples in order to properly interpret test results.

Storage, Handling, and Moving

Soil chemistry changes over time, as biological and chemical processes break down or combine compounds over time. These processes change once the soil is removed from its natural ecosystem (flora and fauna that penetrate the sampled area) and environment (temperature, moisture, and solar light/radiation cycles). As a result, the chemical composition analysis accuracy can be improved if the soil is analysed soon after its extraction — usually within a relative time period of 24 hours. The chemical changes in the soil can be slowed during storage and transportation by freezing it. Air drying can also preserve the soil sample for many months.

Soil Testing

Soil testing is often performed by commercial labs that offer a variety of tests, targeting groups of compounds and minerals. The advantages associated with local lab is that they are familiar with the chemistry of the soil in the area where the sample was taken. This enables technicians to recommend the tests that are most likely to reveal useful information.

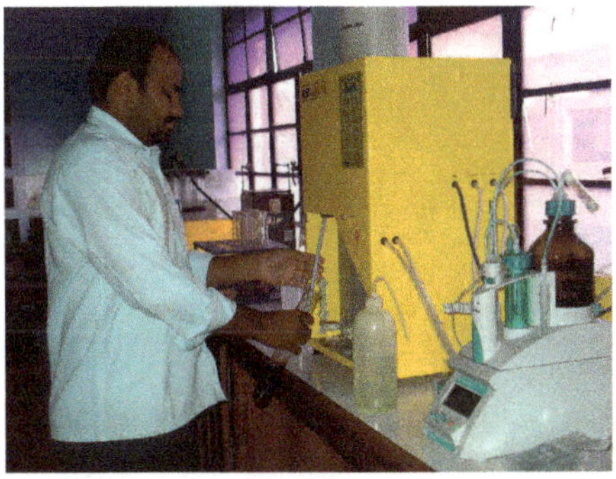

Soil testing in Progress

Laboratory tests often check for plant nutrients in three categories:

- Major nutrients: nitrogen (N), phosphorus (P), and potassium (K)
- Secondary nutrients: sulfur, calcium, magnesium
- Minor nutrients: iron, manganese, copper, zinc, boron, molybdenum, chlorine

Do-it-yourself kits usually only test for the three "major nutrients", and for soil acidity or pH level. Do-it-yourself kits are often sold at farming cooperatives, university labs, private labs, and some hardware and gardening stores. Electrical meters that measure pH, water content, and sometimes nutrient content of the soil are also available at many hardware stores. Laboratory tests are more accurate than tests with do-it-yourself kits and electrical meters. Here is an example soil sample report from one laboratory.

Soil testing is used to facilitate fertilizer composition and dosage selection for land employed in both agricultural and horticultural industries.

Prepaid mail-in kits for soil and ground water testing are available to facilitate the packaging and delivery of samples to a laboratory. Similarly, in 2004, laboratories began providing fertilizer recommendations along with the soil composition report.

Lab tests are more accurate, though both types are useful. In addition, lab tests frequently include professional interpretation of results and recommendations. Always refer to all proviso statements included in a lab report as they may outline any anomalies, exceptions, and shortcomings in the sampling and/or analytical process/results.

Some laboratories analyze for all 13 mineral nutrients and a dozen non-essential, potentially toxic minerals utilizing the "universal soil extractant" (ammonium bicarbonate DTPA).

Soil Contaminants

Common mineral soil contaminants include arsenic, barium, cadmium, copper, mercury, lead, and zinc.

Lead is a particularly dangerous soil component. The following table from the University of Minnesota categorizes typical soil concentration levels and their associated health risks.

Children and pregnant women should avoid contact with soil estimated total lead levels above 300 ppm		
Lead Level	**Extracted lead (ppm)**	**Estimated total lead (ppm)**
Low	<43	<500
Medium	43-126	500-1000
High	126-480	1000-3000
Very high	>480	>3000

Six gardening practices to reduce the lead risk

- Locate gardens away from old painted structures and heavily traveled roads
- Give planting preferences to fruiting crops (tomatoes, squash, peas, sunflowers, corn, etc.)
- Incorporate organic materials such as finished compost, humus, and peat moss
- Lime soil as recommended by soil test (pH 6.5 minimizes lead availability)
- Discard old and outer leaves before eating leafy vegetables; peel root crops; wash all produce
- Keep dust to a minimum by maintaining a mulched and/or moist soil surface

REFERENCES

1. Bonn2011, Conference. "Sustainable Soil Governance: Towards Integrated Management for Water and food Security" (PDF).
2. "Global Soil Partnership". FAO. Retrieved 15 May 2014.
3. Commission of the European Communities. "Thematic Strategy for Soil Protection" (PDF).
4. Follet, R.F (2001). "Soil management concepts and carbon sequestration". Soil Tillage and Research. 61 (1-2): 77–92. doi:10.1016/s0167-1987(01)00180-5.
5. Global Soil Map. "About the Project".
6. Food and Agriculture Organisation, FAO. "World Soil Charter".
7. Lal, R.; Hall, G.F. Miller, F.P (1989). "Soil Degradation: Basic Processes". Land Degradation and Development. 1 (1): 51–69. doi:10.1002/ldr.3400010106. Cite uses deprecated parameter |coauthors= (help)
8. "FAO/UNESCO Soil Map of the World". FAO. Retrieved 22 May 2014.
9. IUSS. "World Soil Day". IUSS. Retrieved 21 May 2014.
10. "International Year of Soils 2015 - IYS 2015". FAO. Retrieved 21 May 2014.
11. Myer, W.B; Turner, B.L (1992). "Human Population Growth and global Land-Use/ Cover Change.". Annual Review of Ecology and Systematics. 23: 39–61. doi:10.1146/annurev.ecolsys.23.1.39.
12. Eswaran, H; Lal,R., Reich P.F (2001). "Land Degradation and Overview".
13. Urban Soil Management Strategy. "Urban Soil Protection".
14. BrownfieldBriefing. "Global Soil Partnership Launched, 19 September 2011".
15. Food and Agricultural Organisation, FAO. "Global Soil Partnership".
16. GSP Technical Working Group. "Global Soil Partnership Background Paper" (PDF). FAO. Retrieved 21 May 2014.
17. "Regional Soil Partnerships". FAO. Retrieved 22 May 2014.
18. "Report of the first meeting of the plenary assembly of the Global Soil Partnership" (PDF). FAO. Retrieved 22 May 2014.
19. "Mandate and Rules of Procedures". FAO. Retrieved 21 May 2014.
20. "Intergovernmental Technical Panel on Soils (ITPS)". FAO. Retrieved 22 May 2014.
21. "The 5 pillars of action". FAO.
22. "Pillar 1: Promote sustainable management of soil resources for soil protection, conservation and sustainable productivity". FAO.
23. "Pillar 2: Encourage investment, technical cooperation, policy, education, awareness and extension in soil". FAO.
24. "Pillar 3: Promote targeted soil research and development focusing on identified gaps, priorities and synergies with related productive, environmental and social development actions". FAO.

25. "Pillar 4: Enhance the quantity and quality of soil data and information: data collection (generation), analysis, validation, reporting, monitoring and integration with other disciplines". FAO.
26. "Pillar 5: Harmonization of methods, measurements and indicator for the sustainable management and protection of soil resources". FAO.
27. Lindbo, David. "The International Year of Soils". Soil Science Society of America. Retrieved 15 May 2014.
28. "Draft resolution submitted by the Vice-Chair of the Committee, Ms. Farrah Brown (Jamaica), on the basis of informal consultations on draft resolution A/C.2/68/L.21" (PDF). United Nations General Assembly. Retrieved 22 May 2014.
29. "International Year of Soils 2015 - IYS 2015". Retrieved 27 August 2014.
30. Hey, C. "EU Environmental Policies: A short history of the policy strategies" (PDF).
31. United Nations. "United Nations (1997) UN Conference on Environment and Development (1992)".
32. Knill, C; Liefferink, D. (2007). Environmental Politics in the European Union: policy-making, implementation and patterns of multi-level governance. Manchester: Manchester University Press.
33. Goulb, J (1998). New instruments for environmental policy in the EU. London: Routledge.
34. Commission of the European Communities. "Proposal for a Directive of the European Parliament and of the council establishing a framework for the protection of soil and amending Directive" (PDF).
35. Europa. "Summaries of EU legislation".
36. Department for Environment and Rural Affairs. "Proposed EU Soil framework Directive".
37. Planning Commission, Government of India (2008). Eleventh Five Year Plan Volume III Agriculture, Rural development, Industry, Services, and Physical infrastructure. New Delhi: Oxford University Press.
38. Press Trust of India. "Gujarat Farmers to get Soil Health Cards".
39. Government of Gujarat. "Soil Health Card Programme" (PDF).
40. Yojana, S. "Soil Health Card Scheme aiming and praiseworthy increase in income of farmers".
41. Chaterjee, K; Chaterjee,A. Das, S. (2009). "India Community Adaptation to Drought in Rajasthan". IDS Bulletin. 36 (4): 33–52. Cite uses deprecated parameter |coauthors= (help)
42. Huq, S; Yamin, F., Rahman, A., Chatterjee, A.,Yang, X., Wade, S., Orindi, V., Chigwada (2009). "Linking Climate Adaptation and Development: A Synthesis of Six Case Studies from Asia and Africa". IDS Bulletin. 36 (4): 117–122. doi:10.1111/j.1759-5436.2005.tb00238. Cite uses deprecated parameter |coauthors= (help)

43. Ministry of Environment and Forests. "Nation action Programme to Combat Desertification- Status of Desertification" (PDF).

44. Ministry of Environment and Forests. "Third National Report on Implementation of United Nations Convention to Combat Desertification" (PDF).

45. Pavao-Zuckerman, D (2008). "The Nature of Urban Soils and Their Role in Ecological Restoration in Cities". Restoration Ecology. 16 (4): 642–649. doi:10.1111/j.1526-100X.2008.00486.x.

46. De-Kimp, C; Jean-Louis, M (2000). "Urban Soil Management: A growing concern". Soil Science. 165 (1): 31–40. doi:10.1097/00010694-200001000-00005.

47. Hossack, I; Robertson, D.; Tucker, P.; Fyfe, C (2004). "A GIS and web based decision support tool for the management of urban soils". Cybernetics and Systems. 35: 499–509. doi:10.1080/01969720490451878.

48. Lal, R (2000). "Soil Management in Developing Countries". Soil Science. 165 (1): 57–72. doi:10.1097/00010694-200001000-00008.

Index